CHASING
CAPTAIN
AMERICA

HOW ADVANCES IN SCIENCE, ENGINEERING, AND BIOTECHNOLOGY WILL PRODUCE A SUPERHUMAN

E. PAUL ZEHR PhD

Published by ECW Press
665 Gerrard Street East
Toronto, Ontario, Canada M4M 1Y2
416-694-3348 / info@ecwpress.com

Cover design: David A. Gee

Library and Archives Canada
Cataloguing in Publication

Zehr, E. Paul, author

Chasing Captain America : how advances in science,
engineering, and biotechnology will produce a
superhuman / E. Paul Sehr.

Issued in print and electronic formats.
ISBN 978-1-77041-199-9 (softcover)
ALSO ISSUED AS: 978-1-77305-141-3 (PDF)
978-1-77305-140-6 (ePUB)

1. Genetic engineering—Popular works.
2. Biotechnology—Popular works.
3. Human biology—Popular works.
4. Bioengineering—Popular works.

I. TITLE.

QH442.Z44 2018   660.6'5
C2017-906591-2   C2017-906592-0

The publication of *Chasing Captain America* has been generously supported by the Government of Canada
through the Canada Book Fund. *Ce livre est financé en partie par le gouvernement du Canada.* We also acknowledge
the contribution of the Government of Ontario through the Ontario Book Publishing Tax Credit and the
Ontario Media Development Corporation.

PRINTED AND BOUND IN CANADA   PRINTING: NORECOB   5   4   3   2   1

Foreword

By Simon Whitfield, Olympic Triathlon Champion

I

Prelude

The Origin of the First Avenger . . .

5

(1) News Flash

Captain America Comes In from the Cold

10

(2) Superhero Science Project

Sowing the Seeds for a Super Soldier Serum

16

(3) Human!

Can Captain America Overcome the Endangered Species inside Each of Us?

27

(4) Shape!

Can We A.I.M. to Make the Star-Spangled Avenger?

43

(5) Muscles in Motion!

Stem Cells, Steroids, and the Sentinel of Liberty

60

(6) Think!

Putting Kapow and Know-How into Cap's Cranium

74

(7) Longevity!

The Steve Rogers Regeneration and Retirement Project

103

(8) Creating Captain America

Engineering a Super Soldier with Sex, Drugs, and Rock 'n' Roll

120

(9) Behold the Future!

What's to Come for Captain America?

145

(10) Pre-Evolving Humanity for Future Frontiers

Bioengineered Superheroes in Space

154

(11) The Ethical Implications of Captain America

Are We Obliged to Enhance Ourselves?

168

Afterword

By Nicole Stott, Artist and NASA Astronaut

181

Acknowledgments

187

Bibliography

189

Index

205

Man is something to be overcome.

—Friedrich Nietzsche, *Thus Spoke Zarathustra*

To me, Captain America symbolizes justice, diligence,
humility, strength, leadership, and a heart of gold . . .

—Sara Gruenwald in *Captain America*
75th Anniversary Magazine #1 (June 2016)

FOREWORD

When I think of Dr. E. Paul Zehr, the image that comes to mind is that of a monk standing patiently at the gate to a monastery, in his hand a key to the great library that lies beyond the garden and the dojo. I don't know if Paul quite understands the profound impact he had on my life when he invited me to walk through that gate and explore the wealth of knowledge with which he engages daily.

I met Paul in 2015, when he invited me to talk to his university students about my experience around performance decision-making and my perspective on the art of mastery. I arrived early and sat in on his class, curious about who this author, neuroscientist, and university professor was. I was immediately captivated by his presentation, the graphic art images he used, his passionate explanations, and his complete grasp of the concepts and ideas he was imparting to his enthralled class. The powerful image of the monk, a superhero in his own right, conveying the human potential with such clarity and insight, began here. I could have sat and listened for hours. I remember wishing I could teleport Paul back in time to meet a younger version of myself and exert his grand master–like influence on my sports career.

I now read everything Paul recommends, and of course have read all of his own books, which have helped me continue to expand my understanding of our vast human potential, from the smallest details to the grand vision to which this book, *Chasing Captain America: How Advances in Science, Engineering, and Biotechnology will Produce a Superhuman*, speaks.

Reading Paul's writing about his Stan Lee correspondence (and the awe to which he assigns a key inspiration), elicits a familiar feeling (of obsession) and a sense that the spark that ignites our focus chooses us—we do not choose it. Much as endurance challenges chose me, superheroes and the art of their portrayal in graphic comics chose Paul. Whether he is in front of his class giving a lecture or sitting in a café telling stories, when he is given a chance to talk about his favorite subject, his chest grows like Superman's, the air crackles as his Spidey-Sense engages, and the subtle nuances of his movements seem to replicate what you might expect if you were in the presence of Bruce Wayne. Reading Paul's books, you can immediately feel passion radiating from the pages, his enthusiasm and passion for life lessons and the universal meanings he sees hidden, like riddles ready to be deciphered, in paperbacks and comic books.

I read *Chasing Captain America* timidly. To be honest, the idea of scientists engineering a "better" us scares me a bit. But Paul's ability to explain complex topics in a simple, concise, and illustrative manner works almost too well. If you are a little afraid of what's coming down the line, like I am, he does too good a job. "Futurist" books often feel farfetched and unlikely, detached from a relatable reality. This isn't the case with *Chasing Captain America*. Paul is not speculating as much as he is laying out the track based on facts, and using the metaphors and imagery embedded in the idea of a superhuman to illustrate the big picture and the inevitable. He understands the science and is able to see the path ahead, illuminated by his vast knowledge of the subject.

Chasing Captain America is full of unique anecdotes and stories, along with thorough projections of the possible unintended consequences of genetic manipulation and enhancement. Paul is not afraid to ask tough questions. For example, we will soon be able to pick and choose our genetic endowments—or rather, they will be chosen for future generations—and Paul asks, Can we really delay the inevitable, and do we even have time to wait? What are the consequences of artificial enhancements? Will society shun "amped" humans by segregating them in an attempt to quell their apparent advantages, or will we embrace them? Is this simply evolution? How far are we willing to go? What are the boundaries, and how will we enforce them? Tough questions that paint a complicated picture, which in turn is thoroughly and effectively explained.

Chasing Captain America tackles the ethical dilemma around the question, "What is human?" Are we responsible for the consequences, or does our responsibility for our current circumstances trump speculation as to what could happen? Are we humans responsible for the continuation of our species? Is there a "bridge too far," or is our evolution to the point at which everyone is a potential superhero inevitable? Are we obligated to enhance ourselves? With the turn of every page, I found myself lost in internal debate while at the same time appraising my new knowledge of how we humans evolved and how we work. *Chasing Captain America* paints a fascinating picture, full of both hope and the potential for despair. We ignore it at our peril.

SIMON WHITFIELD, Olympic Triathlon Champion

PRELUDE
THE ORIGIN OF
THE FIRST AVENGER . . .

The late 1930s and early 1940s witnessed the birth of the super-hero comic book. Action Comics showed us "Superman," Detective Comics debuted the "Bat-Man," and Timely Comics (which later became Marvel Comics) brought us "Captain America." Joe Simon (1913–2011) and Jack Kirby (1917–1994) were the original powerhouse tandem in this golden age of comics. Together they cocreated the "Star-Spangled Avenger" in his debut in *Captain America Comics* #1 in March 1941. The story was naturally entitled "Case No. 1: Meet Captain America," written and drawn by Joe Simon and Jack Kirby. We weren't introduced to him as Steve Rogers yet, but a key panel says that "a frail young man steps into the laboratory" and is "inoculated with the strange seething liquid" by Professor Reinstein. We can see the reaction of the observers and the growth of Steve Rogers in the panel shown in Figure 1.

In Reinstein's words, the liquid "formula" will move "through his blood . . . rapidly building his body and brain tissues, until his

FIGURE 1: The first glimpse of the "super soldier" procedure that produced Captain America in March 1941. But what could have been in Professor Reinstein's serum? (*Captain America* © Marvel Comics Inc.)

stature and intelligence increase to an amazing degree." As drawn in the panels, Rogers appears to grow about 18 inches taller and double in size to become "the first of a corps of super-agents whose mental and physical ability will make them a terror to spies and saboteurs. . . . We shall call you Captain America, Son!" The injection of the synthetic formula brings about a change—growth—in Steve Rogers within minutes. This seems impossible, but what would really be in such a formula, and could it work? The answers lie in the pages that follow.

It was during the early days of Captain America that Stan Lee—quintessential comic book icon—made his debut. Lee's first writing credit was the May 1941 story "Captain America Foils the Traitor's Revenge" in *Captain America Comics* #3, the first of many, many amazing stories. Stan Lee spearheaded the creation of the pillars of the Marvel comic book canon, including Spider-Man, the Hulk, Doctor Strange, the Fantastic Four, Iron Man, Daredevil, Thor, and the X-Men.

Stan Lee also brought back Captain America as a Marvel hero. In March 1963, in *Avengers* #4, Stan Lee and Jack Kirby revived the Cap from his deep freeze. His origin story, another Lee and Kirby collaboration, came later in "The Hero That

Was," which appeared in *Captain America* #109 in January 1969. The retelling of Cap's beginnings—see also another version in "The Monster Unmasked" flashback in *Captain America* #100 (April 1968)—comes from Steve reminiscing with Nick Fury: "I remember how it all began! It seemed like any other day to me then—how could I know—how could I have even suspected—that I was about to be reborn!"

We learn that Steve was "the skinniest, scrawniest recruit ever to report for induction," and that he was repeatedly declared "too frail for military duty!" But apparently he was perfect for participation as "a human guinea pig in a deadly experiment." Spindly Steve was then introduced to "Professor Reinstein—the world's greatest physicist." We see the transformation in the panel in Figure 2.

In a critical run in the *Captain America* issues 444–54, writer Mark Waid and artist Ron Garney tell what happens to Captain America when the Super Soldier Serum "wears off." This builds on the story arc that Mark Gruenwald used to end his 10-year run, which concluded with the failure of the serum and Cap's apparent death. In Waid's critical narrative, Captain America's demise is followed by a resurrection, using serum from his nemesis, the Red Skull. This was part of an arc called "Operation Rebirth" and hints at the fluidity of the adaptation that gave Cap his superpowers. But is it really possible for a physiological adaptation to "wear off"? Yes? No? Read on to find out!

Like all comic book superheroes (especially those who've been around for a long time), Cap's origin story has been tweaked, revised, and rewritten numerous times. This re-visioning process also affects the big-screen Captain America, played by the incomparable Chris Evans. Cap's Marvel Studios backstory, seen in *Captain America: The First Avenger* and the basis for the character in *The Avengers* and *Captain America: The Winter Soldier*, is

FIGURE 2: Stan Lee and Jack Kirby's reimagining of the super soldier procedure in *Captain America* #109 1969. Lee and Kirby add the "Vita-Ray" treatment, but what would this really be? (© Marvel Comics Inc.)

heavily influenced by the plotlines found in the Marvel Universe that holds the "Ultimates." In the Ultimates universe inhabited by Mark Millar and Bryan Hitch, the U.S. military used a pharmacological cocktail found in Dr. Abraham Erskine's (the original name of Professor Reinstein) "Super Soldier Serum" and "Vita-Ray" treatment to transform the slightly built soldier (or actual soldier reject) Steve Rogers into the "perfect" specimen of human development and conditioning. As must now be clear, he really is the ultimate example of genetic engineering and science. But what exactly would need to be done to build Captain America? And what would be in the Vita-Ray?

In answering these questions and writing this book, I leaned heavily on my science background in kinesiology and neuroscience while also reaching out to scientific and biomedical experts around the world. This journey took me from the panel in Figure 1 of the 1941 *Captain America* comic book to the near future exploration of outer space. The story you are reading is about the real science behind the Super Soldier Serum and the transformation of Steve Rogers from slightly built human into superhuman. Here we explore, examine, and answer what the real super soldier treatment might have been and might one day become. And whether we should pursue such a program at all.

1. NEWS FLASH
CAPTAIN AMERICA COMES
IN FROM THE COLD

Any hope of reproducing the program is
locked in your genetic code . . .
—Agent Peggy Carter to Steve Rogers,
in *Captain America: The First Avenger*

Project Super Soldier is a success!
I know it! I can feel it!
—Steve Rogers, minutes after growing into his newly
enhanced body, in *Captain America* #1

One hundred trillion cells. One hundred thousand kilometers of
blood vessels. Six hundred and forty muscles. Seventy kilometers of nerves. Eight meters of intestines. And at least five vital
organs. These make a human. And they were all supercharged
with the birth of Captain America.

Captain America came to be because of efforts to create an
enhanced soldier during World War II. As Colonel Chester

Phillips (played by Tommy Lee Jones) told us in the 2011 Marvel Studios film *Captain America: The First Avenger*, "The Strategic Science Reserve is an allied effort made up of the best brains in the world. . . . Our goal is to create the best army in history. . . . It starts with one man. He will be the first of a new breed of super soldier."

Steve Rogers—the man who would become the first super soldier—was an unlikely candidate for super status. Colonel Phillips, speaking with the lead scientist on the project, Dr. Abraham Erskine (aka Professor Reinstein in many of the comics), remarks, "You're not really thinking about picking Rogers, are you? When you brought a ninety-pound asthmatic onto my base I let it slide. . . . Stick a needle in that arm, it's gonna go right through him . . ." But Dr. Erskine argues that Rogers is "the clear choice. . . . I am looking for qualities beyond the physical." Clearly Steve Rogers had something more, which he demonstrates when he dives onto a grenade (part of a test) to save his platoon and everyone around him.

The characteristics that set Rogers apart appealed to me when I started reading comic books as a kid. I wasn't the first comic book reader in the family. My mom loved them, too, when she was growing up in the 1930s and '40s. I wish she had kept some of the Action and Detective Comics she must have had. Anyway, my mom would say "reading is reading," and she would bring home a bunch of comics for me when she went grocery shopping. I grew up in a small, small town (I'm looking at you, Chesley, Ontario!). So, yes, the grocery store was also the "book" store.

There were many comic book characters that interested me back then: Batman, Iron Man, Daredevil, Thor, Nova, the Flash, and a host of others—including Aquaman (please don't judge me). Above all, though, I loved reading the "team-up" books like the Fantastic Four, the Justice League, the Justice Society, and even the Defenders. But the all-time team-up for me was always *The Avengers: Earth's Mightiest Heroes*. Most of all, I admired Captain America.

At first I was mostly captivated by his awesome red, white, and blue outfit and amazing indestructible shield. Only later did I learn something of his history. Captain America goes way back to the golden age of comics. In his first appearance in *Captain America Comics* #1 in March 1941, Captain America was shown as a tactically savvy, strong, and smart fighter; an overall amazingly conditioned human; and a fearless and principled leader.

Like Batman, Captain America has had a long career as a superhero. Unlike Batman, who attained his abilities through arduous training, Captain America's powers came from a special Super Soldier Serum courtesy of Uncle Sam. With Captain America, we take the idea of a human superhero and bend biology.

Also like Batman, Captain America had a noble mission. In Cap's case, it was to fight tyranny and oppression in the war. As the voice-over informed the audience in the 2014 Marvel Studios film *Captain America: The Winter Soldier*, Cap is "a symbol to the nation, a hero to the world . . . the story of Captain America is one of honor, bravery, and sacrifice." Professor Reinstein says that Cap represents "the crowning achievement of all my years of hard work! The first of a corps of super-agents whose mental and physical ability will make them a terror to spies and saboteurs!"

Captain America's origin story has been tweaked, revised, and rewritten numerous times over the decades. As seen in a number of movies, including *Captain America: The First Avenger*, *The Avengers*, and *Captain America: The Winter Soldier*, his backstory is heavily influenced by the plotlines found in the Ultimates Marvel Universe.

The initial run of Captain America ended with *Captain America* #78 in September 1954. Later, Stan Lee brought him back to with the Avengers. The original lineup in *The Avengers*

\#1 in September 1963 included Iron Man, Ant-Man, Wasp, Thor, and the Hulk. Although he has been heavily identified with them, Captain America didn't actually appear until issue \#4. No offense to the Hulk, Ant-Man, or Wasp (all part of the "Founding Members" group), but they put the damper on how cool the Avengers could be. By adding Captain America to the fold, along with Iron Man and Thor as leads, Lee created a nearly perfect team. With this lineup, how could you go wrong?

CREATING CAPTAIN AMERICA
ENHANCING EVOLUTION WITH ENGINEERING AND SCIENCE

This book is organized around the theme of human biological adaptability in light of our ever-expanding scientific and technological prowess. We explore the ability we now possess to change the natural function and capacity of the human body. How can we create a superhuman (something that's also been called meta-human, mutant, mutate, cyborg, posthuman, and so on in comic books, movies, and transhumanist literature)?

The idea that there can, or might, be humans with extraordinary powers is not new. Popular models have always been with us. In days gone by, they were found in Greek, Roman, and Norse mythology, as in the exploits of Hercules, Mars, and Thor. They live on in the guise of modern-day superheroes. We read about them in comic books and see them in movies and television shows. We also see superhuman performances in the sports arena.

We will touch on this history while also taking an inside-out tour of the human body, from what we can see with the naked eye to the things we can change inside. We will delve deep down to the interior of our cells, into the nucleus, where lives the very fabric of who we are: our DNA.

Think of the human body as a garden. Think of the history of our species before we even had gardens and just foraged for plants and berries. Over time we learned to plant what we wanted and to trim away things we didn't want. Then we started using fertilizer, pesticides, and chemicals to make the things we wanted grow better and to inhibit the things we didn't want. Then we found ways to go into the plants themselves and change how they worked by changing the DNA of the cells. That's all going on in your garden and, increasingly, in your body. The thing is that now we are both garden and gardener.

In *Chasing Captain America* we explore a number of mythological and pop culture icons, some ancient and some modern, including a few elite athletes. All of them possess a backstory related to the exploitation or manipulation of the human body.

BIOENGINEERING AT ITS BEST

Chasing Captain America explores the modern-day convergence of biology, engineering, and technology. We are moving rapidly toward a destination for our species that is not simply the result of natural selection. The ability to select for ourselves the traits we want is almost within our grasp.

This book completes my loose trilogy of popular science books that began with *Becoming Batman: The Possibility of a Superhero*, which is all about the potential to train the biology of the body. It was followed by an exploration into how technology can be used to amplify biology in *Inventing Iron Man: The Possibility of a Human Machine*. Here the focus is on changing human biology itself.

In *Chasing Captain America* we look at how humans can alter our abilities through surgery, pharmaceutical enhancement, technological fusion, or genetic engineering. And we provide tentative answers to a number of intriguing questions: What are we and what can we become? What are the real limits of being human? How far can those limits bend before we are no longer human? And, probably most importantly, how far should we bend those limits?

2. SUPERHERO SCIENCE PROJECT
SOWING THE SEEDS FOR A
SUPER SOLDIER SERUM

Stem cell therapy, robotic prosthetics, face transplants—it's all the
stuff of science fiction. "Impossible" is being phased out . . .
—Kelly Sue DeConnick, writing in *Captain Marvel* #12

Our destiny is to become like the gods we
once worshipped and feared . . .
—Michio Kaku, *Physics of the Future*

The 1960s were a tumultuous period but also a time of renewal.
Captain America was revived and brought out of the deep
freeze—literally—by Stan Lee in *Avengers* #4 (March 1964).
Later, on March 18, 1968, Robert F. Kennedy spoke at the
University of Kansas. In his speech he quoted George Bernard
Shaw, saying, "Some people see things as they are and say why?
I dream things that never were and say, why not?" This sounds
like something that Stan Lee might have written for one of his
many legendary comic book characters.

Captain America represents a blurring of the lines separating fact and fiction, dreams and reality. Currently our species is at a point where our dreams can scarcely compete with what's happening in the world. For our purposes, Captain America represents the extreme range of adaptability of our species: he represents the idea of actually changing what it means to be human.

Of course, we—as a species—are always changing. The only real constant in life is that constancy is a state of flux. In the words of Giuseppe Tomasi di Lampedusa, from his book *The Leopard*, "If we want things to stay as they are, things will have to change."

In physical anthropology, it is accepted that the divergence between hominids (that's us) and our closest cousins, the chimpanzees, occurred between five and seven million years ago. The famous Laetoli footprints near Olduvai Gorge in Tanzania provide fossil evidence for the emergence of the upright bipedal walking characteristic of our species within the past four million years.

Homo sapiens means "wise man" or "knowing man" in Latin. It's the term used to describe our species as it emerged from Africa some 200,000 years ago. We achieved our current set of characteristics about 50,000 years ago when we acquired complex language and the symbolic thought that goes with it. We have been mostly the same (at least compared to how we rate against other species) since that time. It has been said that it's what's inside that counts, and we've been the same inside for a long time.

Our species, like all others, has slowly and steadily been changing and adapting to our environment. Now, though, the freaky among us—those who are taller, quicker, stronger—are getting freakier faster. Consider "wingspan," a measure of your height compared with the distance between the fingertips of both your outstretched hands. It used to be that for most people wingspan was roughly equivalent to their height. Well, in 2011, a 16-year-old high school basketball superstar, BeeJay Anya, with a height of 6'8", was measured with a wingspan of 7'9". And in March 2017, Andy Staples wrote in *Sports Illustrated* about a high school football–playing giant of a man standing 6'9" tall and weighing in at 396 pounds.

When we think of what "human" superheroes like Batman and Captain America can do, modern-day athletes come to mind. The size and ability of elite athletes are well beyond anything that can be considered "normal." The media provides us with many instances like these of extreme human performance. At the same time, there is an explosion of interest in genetics, technology, and medicine, with significant implications for human life and society—and especially for the future of our species.

The Super Soldier Serum and Vita-Ray treatments employed in the making of Captain America have their modern analogues. We hear a lot about new and more complex surgeries, technological implants, drug therapies, and genetic manipulations. Concern about the possible negative side effects of these innovations is balanced against our age-old impulse to push beyond the normal biological limits of the body. Until recently, these new treatments were mostly things like so-called dietary supplements or steroids. Nowadays, however, drug doping has met gene doping while saying hello to synthetic biology. Entirely new powers for modifying life have been harnessed.

The fantasy worlds of ancient mythology and modern comic book superheroes have all been about exceeding conventional human limitations. Now, however, the line between fantasy and reality is blurring and, in some cases, being erased altogether. The form and function of the human body are now fluid.

WHAT IS IN CAP'S JEANS ISN'T WHAT WAS IN HIS GENES

Biological determinism is a scientific term with a simple meaning. You—I, we, they, all biological organisms—grow and develop based on the natural biological material you're born with. The biological material that produced the spindly army reject Steve Rogers is the deoxyribonucleic acid—the DNA—he got from his parents, Sarah and Joseph.

The environment you grow up in and the experiences you have had, are having, and will have shape the expression of that DNA—meaning who you are, what you can do, and how you react in the environment. But you still just get a certain set of genetic material when you are born. That's your body, and you don't get another. In the case of Steve Rogers, that should have been it. He never would have made it into the army. He certainly wouldn't have evolved into a super soldier.

Fictional Steve Rogers had access to the full-on Super Soldier Serum treatment. In reality, it isn't that simple anymore—we even have XNA. That stands for xeno nucleic acid, a synthetic alternative to the naturally occurring DNA and RNA (ribonucleic acid) you have in your body. We may not, after all, be stuck with the body we were born with.

Back in 1974, Rudolf Jaenisch and Beatrice Mintz demonstrated that DNA could be transferred from one species to another. They created the first genetically modified animal when they inserted a DNA virus into a mouse. More than 30 years later, in July 2010, a U.S. genetics team took the concept a giant step further by creating synthetic DNA. The team, headed by maverick molecular biologist Craig Venter, inserted synthetic DNA into a bacterial cell that subsequently grew and multiplied.

In other words, it worked. And this, in terms of the potential for genetic modification, changes everything.

The importance of this success lies in the fact that the DNA was synthetic—created by human hand. Venter and his team combined the approaches of molecular biology and nanotechnology to deliver a sledgehammer blow to biological determinism. Then, as a further testament to our increasing ability to modify living organisms, in March 2016 Venter's group created a truly minimalist cell—fully functional with only 473 genes.

THE MERRY MARVEL MARCHING SOCIETY STARTS TO SPRINT

Recent advances in biomedical science and engineering have fundamentally altered the potential role we can play in shaping our own biology. It puts into our frail human hands powers— superpowers—that we previously could not hope to possess or harness. This is a staggering, beautiful, elegant scientific result.

It is also Frankenstein as reality if you prefer Mary Shelley, the island of Dr. Moreau if you are an H.G. Wells aficionado, or the Incredible Hulk if you are a fan of Stan Lee. Those imagined creatures and worlds are the metaphors we will use here to explore how science fiction and science fantasy are converging on new science fact.

The current state of science and the endlessly expanding power of biomedical engineering have the potential to fundamentally alter the future of the human species. What superheroic powers lie within our grasp? There may be no limits beyond what we can imagine.

Developments in robotics, natural and synthetic steroids, genetic engineering, and nanotechnology have us on an uncharted course as a species. What we explore in this book is what it means when a species acquires the ability and the desire to change itself. And not simply external change, like wearing clothing or having surgery; I mean internal modification using technological knowledge. We are moving toward a time when it will be possible to modify the structure and function of the human species itself.

POWER OF NATURE AND NURTURE

Captain America's adventures show him battling the "Axis of Evil" in World War II. In the real war, the winter of 1944 was a long, hard experience for the Dutch. So hard, in fact, that the winter got its own name. It was called the *Hongerwinter*—the hunger winter. To punish the Dutch for their resistance to the occupation of their country, the Nazis imposed a blockade, cutting off food and fuel shipments to major cities. The blockade affected the life and health of more than 4.5 million people and led to the death of almost eighteen thousand due to famine.

What I want to discuss, though, are those who survived. And in particular, the women who were pregnant during the hunger winter. How did the famine affect their babies? The event has been carefully studied as an instance of famine in a modern industrial society. When the previously healthy mothers became malnourished, their babies were bound to be at risk of health problems. And, indeed, the children of the pregnant women living through the hunger winter were more susceptible to diabetes, obesity, heart disease, and many other conditions. In fact, babies born during and shortly after the famine were smaller than normal. But there's something more that wasn't so obvious at the time.

When those smaller-than-normal children grew up and had babies of their own, their babies—and keep in mind that this means a generation after those whose mothers were actually exposed to famine—were also smaller than normal. The unavoidable conclusion is that exposure to the famine changed the mothers and their genetic material so that those changes were handed down to the next generation. The environment—nurture—can change nature, and some of these changes alter the genetic expression of the person. This was one of the first true realizations of the emerging field of epigenetics (more later!).

Since the environment sculpts who we are, we may have to deal with some unintended effects of changes in it. Over the last fifty thousand years of our evolution as a species, changes have occurred in our genome that allowed the strongest among us to flourish in harsh environmental conditions. This is "natural selection" as described by Charles Darwin and Alfred Russell Wallace.

But now we have changed our environment so much that it is virtually unrecognizable from even a few hundred years ago. There is no need to gather our own food, make our own clothes, or use our bodies to move from place to place. You could say we used to eat to live but now we live to eat. Can we now rely on the environment to change us by encouraging adaptive evolutionary trends?

ACCELERATING EVOLUTION
GOING BEYOND THE BEAGLE

There was a time when our ability to alter genetics was captured in the phrase "choose your parents wisely." In other words, if you aren't born with it, you'll never have it. Steve Rogers came face to face with this sobering fact when he tried to enlist in the army.

This science of genetics was heavily influenced by Charles Darwin in his 1859 opus *On the Origin of Species by Means of Natural Selection*. Darwin spent five years (1831–36) voyaging around the globe and surveying plant and animal life in South America, Tahiti, and Australia on HMS *Beagle*. The species he discovered and the observations he made about their physical features laid the groundwork for his theory of evolution.

The central thesis of Darwin's theory is that environmental pressures lead to adaptations in organisms that are beneficial to survival of the organism in that environment. Now we can modify our environment to suit our needs and are close to choosing our own genetic makeup. Picking and choosing from the genetic traits our parents gave us is just over the horizon. We may need this ability to survive—we may not have time to wait for the random process of natural selection.

But how do we define evolution anyway? It seems to have many different meanings and connotations depending on the context. Back in 2001, John Wilkins at the National Center for Science Education wrote an article called—wait for it—"Defining Evolution," in which he offered this: "Transmutation (descent with modification): This is the notion that new species emerge from existing species and that all existing species are the product of change in older ones." In this book I use the term *evolution* to refer to functional adaptation of a species to the environment in which it lives.

Again, consider what our species can achieve. For me the convergence of superb and superhero achievement was evident in the men's triathlon at the 2000 Sydney Olympics. Canada's Simon Whitfield crashed his bike and seemed to be out of the race, but he went on to win the triathlon in Olympic record time. This was just one highlight in a career that included 10 consecutive Canadian titles and a silver medal in 2008 at the Beijing Olympics.

Chris Gehrz, professor of history at Bethel University in St. Paul, Minnesota, compared selected men's Olympic track and field records from 1908 to 2008. Gehrz showed huge percentage improvements over that hundred-year span: discus (70.9%), pole vault (60.7%), marathon (26.2%), and 100 m dash (10.3%). Without a doubt we are getting freakishly faster all the time.

Performance improvements and records are achieved by a combination of genetic endowment and extreme training. They underscore the amazing capacity of the human body to adapt to stresses and allow performance at the highest possible level. I am stunned by the scope of human performance abilities that can be unleashed in our species. Achievements that exceed the limits of "human" are on the horizon. But what are human limits and where are we actually headed?

When we talk about regenerative medicine, we're describing the replacement, bioengineering, or regeneration of our cells, tissues, or organs with the aim of restoring "normal" function. The next century holds great promise for regenerative therapies. Milica Radisic is a chemical engineer and professor of biomaterials and biomedical engineering at the University of Toronto. Radisic told me that "regenerative medicine is in the same place that pharmaceuticals were in the 1930s and 1940s— struggling with few profitable products. It is almost as if we need a bigger driving force to turn that corner. In the 1940s there was WWII and a lot of money was poured into pharma to make penicillin with immediate human trials. War efforts brought also the first computer." While she wasn't suggesting that we need a war to advance our species, Radisic's comments do highlight the key issue—that it takes an enormous, orchestrated effort, with a lot of financial investment, to make major biomedical advances.

We may want to move things along faster, but we have to balance scientific progress against profits and losses to avoid making catastrophic mistakes. This means carefully translating results from trials in other animals before testing in humans, and no less carefully studying possible different reactions among children and adults, men and women, and other groups. Radisic explained that "assistive technologies that are now being developed for disabled individuals—including exoskeletons—will enter into the mainstream, they will be as commonly used as cell phones or internet are used today." She cautioned that "loss of privacy will come as an unwanted side effect."

When asked about how her work might be misused, Radisic pointed out that "it is already happening. We don't have to wait for the next century to see the negative extrapolations of our work with all of the fake stem cell treatments out there . . ." In order to move forward, Radisic believes that the more informed we are about the possibilities and limitations of biomedical science, the better off we will be.

A particularly exciting part of Milica Radisic's research is on using regenerative medicine to repair heart cells—cardiomyocytes—to help recovery in cardiac disease. Since enhanced heart muscle function could also provide a performance enhancement in an otherwise healthy person, I asked her about the application of her work to augmentation and enhancement in athletics or other scenarios. She warned that this would have "to be done very carefully, and one of the biggest negative outcomes could be a non-selective augmentation."

"Our bodies develop with pieces put together to function in a certain harmonious order," continued Radisic. "This is true for most people, although some are born with birth defects and for them, regenerative therapies would presumably be appropriate. However, what happens to that individual if you give a normally functioning individual an athlete's heart without also enhancing their circulatory system, their nervous system (for improved reflexes), their skeleton, their muscles, etc.? What if we chose to enhance everything in their body? We can glimpse part of the answer in people who go overboard with cosmetic surgeries. It is not always a pretty outcome."

Milica Radisic's work challenges us to reassess what constitutes "normal" and "natural" in the range of human abilities. We once used superheroes as foils for our dreams. Now we are prepared to move beyond dreams to a new reality. But we're going to begin to examine that move by first trying to understand what "being human" with "normal ability" really means.

3. HUMAN!
CAN CAPTAIN AMERICA OVERCOME THE ENDANGERED SPECIES INSIDE EACH OF US?

_____

The changes crept in around the edges, too slow to be noticed,
like mold on bread. Fixing serious medical problems first but
always moving closer to the simple trials of daily life.
—Daniel H. Wilson, *Amped*

Notice how balance and coordination are more
important than mere brute strength! The important
thing is to be agile, alert, ready for anything!
—Captain America, in *Avengers* #5

I doubt that Steve Rogers is a hard rock fan. Usually he's shown as having the musical sensibilities you'd expect for someone growing up in the 1930s and 40s. While I regularly listen to big band music—Benny Goodman and "Sing, Sing, Sing," are fantastic—I am a hard rock fan at heart. In May 1983, the British heavy metal band Iron Maiden released a hugely influential and relevant album—*Piece of Mind*. The third song on that album, "Flight of Icarus," has always been a favorite of mine.

The Greek myth of Icarus is a cautionary tale about human ambition and hubris. Daedalus builds feather-and-wax wings to be used by his son Icarus to escape the island of Crete. His father warns him not to fly too high—not, as in the Iron Maiden lyrics, to "fly as high as the sun." Yet, overcome by the feeling and power he experiences in flight, Icarus does exactly that. The heat of the sun melts the waxen wings, and Icarus plummets to his death in the Aegean Sea.

In his book *The Big Bam: The Life and Times of Babe Ruth*, Leigh Montville writes that "the American hero was constructed differently from the classic heroes of the Greeks and Romans. Virtue never had been a necessity." This may sometimes be true, but Captain America is virtue personified. As Mark D. White writes in *The Virtues of Captain America: Modern-Day Lessons on Character from a World War II Superhero*, Cap "provides an example of the personal virtues that philosophers since ancient times have put forward as defining personal excellence." Does Captain America know better than Icarus to never fly too close to the sun?

As a species, we use scientific advances to change who and what we are. It's as though our constantly improving technologies have an unstoppable momentum of their own. As with computing power and smartphones, we always want more. What was heralded as the next big thing and celebrated as a marvel quickly falls out of style and becomes last year's trend.

It makes sense that we would reach for more—like Icarus, that we would fly higher—when it comes to enhancing human ability too. The question is, at what point have we gone too far beyond normal? Is there such a point? In the past, we have turned to fictional superheroes to explore these questions. Until recently that's all we could do.

Now, though, we have the power to change the very fabric of life. We are obliged to ask ourselves, what does it really mean to be human?

What's "natural" or "normal" has always been a moving target, a pendulum swinging back and forth in front of us. Now the pendulum is moving much faster and taking much wider swings. Do we have an adequate definition for the accepted range of "normal" human ability? (Spoiler alert—a valid concept of normal has never really existed.) In this context, how constrained are we by traditional thinking? In the words of futurist, novelist, and intellectual agent Warren Ellis in his graphic novel *Transmetropolitan, Vol. 4: The New Scum*, "tradition is one of those words conservative people use as a shortcut to thinking."

Let's consider the traditional boundaries that define our societal understanding of what is acceptable "normal" function and consider the ethics of neural enhancement.

TEST BOUNDARY
DO PROSTHETICS RESTORE OR ENHANCE HUMAN ABILITY?

Captain America's sidekick, introduced as a boy in *Captain America Comics* #1 (March 1941), disappears, seemingly killed in action in World War II. However, he reappears as the "Winter Soldier"—a brainwashed Russian/Hydra operative periodically "frozen" and then resuscitated for missions. Crucially, Bucky as the Winter Soldier sports an amazing bionic arm. This gives him powerfully enhanced ability.

The issue of enhancement has been tested in the sports arena too. The most high-profile (and ultimately tragic) example is that of South African sprinter Oscar Pistorius. As I discussed in *Inventing Iron Man*, when he was a competitive sprinter, Pistorius was known as "the fastest man on no legs" because he is a double below-knee amputee.

Pistorius was born with fibular hemimelia (longitudinal fibular deficiency), which means that he lacked the long bones on the outside of his shins. Because of this condition, the chances of Pistorius being able to stand and walk were considered very low. Accordingly, his parents made the difficult decision when he was about 11 months old to allow surgical amputation of his lower legs. Functionally, this bilateral leg amputation allowed Pistorius to use prosthetics that could carry his weight. He used more or less conventional prosthetics for walking. Later, for running he used "blades," so-named because the carbon-fiber prosthetic, the Össur Flex-Foot Cheetah, is shaped like a blade.

The combination of his prosthetic limbs and fantastic sprinting biomechanics created considerable controversy when Pistorius competed in an International Association of Athletic Federations (IAAF) event held in Rome in 2007. Some observers suggested that his blades actually gave him a performance advantage over able-bodied runners using two legs. That is, the prosthetic technology allowed him to exceed human limitations.

In 2007 the IAAF introduced Rule 144.2, which prohibits "any technical device incorporating springs . . . that provides the user with an advantage over another athlete not using such a device." This prohibition ended Pistorius's quest to compete in the 2008 Olympics.

Scientific analysis suggested that the carbon-fiber blades significantly improved the runner's mechanical efficiency. Since the blades are actually much lighter than the lower legs they replaced, they can be moved about 15 percent faster than the legs of even the most proficient elite athletes (including, for example, Olympic gold medalist Usain Bolt of Jamaica). The blades also required 20 percent less force than able-bodied runners needed to achieve similar speeds.

However, in May 2008 the Court of Arbitration for Sport based in Lausanne, Switzerland, reversed this decision. This court ruled that the IAAF had failed to prove that the blades provided an unfair advantage. Legal maneuvering had the IAAF rescind its ban on use of the blades. In the summer of 2011, Pistorius ran a personal best of 45.07 seconds in the 400 meters, thus meeting the "A" standard 45.25 seconds. This result got him on the South African team for the 2011 IAAF World Championships in Daegu, Korea. In 2012, Pistorius was selected to compete for South Africa in the 400 meter and 400 meter relay at the London Summer Olympic Games. In so doing, Oscar Pistorius became the first amputee runner to compete at the Olympics. The South African relay team finished next to last in the final, but Pistorius's achievement was recognized when he was given the honor in the closing ceremonies of carrying the flag for his country. This left audiences with the perplexing paradox of someone with an obvious physical disability—no lower legs—running faster than most able-bodied runners by using special equipment.

Brendan Burkett, professor of biomechanics at the University of the Sunshine Coast in Australia, raised an interesting question in this regard. If technology can amplify human performance and alter outcomes, how do we ensure equal access for all? The winner of the marathon at the 1960 Rome Olympics, Abebe Bikila of Ethiopia, ran the entire race barefoot. What would his time have been if he had worn technically advanced running shoes?

The examples of Pistorius and Bikila suggest in different

ways a fascinating possibility: that a physical impediment to performance could become the basis for performance enhancement. One can take this a speculative step further: suppose a talented runner with "normal" legs who wants to undergo surgical procedures to remove their lower legs so they can be fitted with a performance-enhancing prosthetic. What do we as a society say then? Is that "normal"? Almost certainly it's time to phase out the concept of "normal" altogether.

The takeaway here is that amplifying or enhancing human ability carries with it some ethical and societal implications. This is obviously the case for even more "conventional" mechanical prostheses like that in Pistorius's situation. This needs thinking about as we continue to move toward more technologically complex and integrated prosthetics such as brain–machine interfaces and other implanted stimulators.

AN IMMINENT BUT UNTESTED BOUNDARY
NEURAL ENHANCEMENT BEYOND THE "NATURAL" FUNCTIONAL LEVEL IN "NORMAL" HUMANS

We have seen that conventional mechanically driven technology, originally designed for restoration of function, can also be applied to enhance "normal" performance. Where is the boundary separating acceptable from unacceptable enhancement? And what happens to our perception of that boundary when we use technology to change the human inside as well as out?

Let's consider, for example, what we can achieve by stimulating brain function. Normal brain function is due to the electrical activity between neurons. It can be altered by outside intervention. In fact, people have been directly stimulating brains since Gustav Fritsch and Eduard Hitzig electrically mapped a dog's brain in 1864. Until recently, though, brain stimulation has

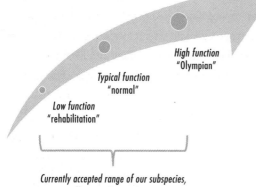

What does it mean
to be here?

High function
"Olympian"

Typical function
"normal"

Low function
"rehabilitation"

Currently accepted range of our subspecies,
Homo sapiens sapiens

FIGURE 3: Continuum of human performance abilities from low to high function.

remained largely in the realm of neuroscience research and some clinical diagnostics.

Now, advances in and access to technology have spurred interest in repetitive transcranial magnetic stimulation (rTMS). This technique makes use of electromagnetic induction to activate brain cells from the surface of the skull. It had its debut as a clinical and research tool, but rTMS is now being used for treatment of chronic pain syndrome, depression, Parkinson's disease, personality disorder, post-traumatic stress disorder, stroke, and bipolar disorder, and for enhancing motor learning. Most of these are clinical examples that would be in the restoration or "rehabilitation" range, as outlined in Figure 3.

Yet it is not a big step to use the same technique to enhance attention and improve performance in those who already operate in the "normal" range shown in Figure 3. Indeed, Vincent Clark and Raja Parasuraman in a 2014 editorial on enhancing brain function wrote that rTMS and related brain stimulation method-ologies "can be used to improve attention, perception, memory and other forms of cognition in healthy individuals." Others have shown that transcranial direct current stimulation could be used to enhance vigilance in people with no deficits in that realm. There are now even commercially available devices that may be able to do this.

So now we're asking questions about transhumanism. Is it acceptable in our society for someone to seek a performance advantage by replacing healthy body parts with mechanical ones, or to enhance brain function with technological constructs? Silvia Camporesi, neuroethicist at King's College, London, posed a related question: What are the ethics of enhancement using assis-tive device technology (such as currently available neuropros-thetics) or genetic intervention (which will soon be accessible)? Of course, it becomes much more difficult to evaluate "normal" or "natural" human ability when biology, modern technology, and neuroengineering merge to produce "bionics." Another term, *cybernetics*, is increasingly in play, and more directly suggests a control system combining artificial intelligence and machine-bi-ological interfaces.

From cybernetics it is a very short jump to the term *cyborg*. All of these jumps take us further and further away from the "human" range of ability. They do, though, provide a frame of reference for what society has come to accept. Cyborgs can be seen in many pop-culture references. They figure prominently, for example, in BBC TV's *Doctor Who* series. *Doctor Who*—the longest-running science fiction show in TV history, according to Guinness World Records—features cyborgs in the form of "Cybermen." The Cybermen—and recently Cyberwomen—are discussed in enter-taining detail along with everything else in the Doctor Who universe by Paul Parsons in *The Science of Doctor Who*.

Doctor Who's "Cyberpeople" represent an extreme conception of biological and machine connection. They have a significant biological base, including an artificial nervous system, encased by an iron robotic exoskeleton. The rise of the Cybermen is presented in the TV series as a parallel humanoid species that began implanting technology and artificial parts into their bodies until they one day crossed the threshold separating humanoid species and cyborgs.

So much is fiction—for now. Clearly, though, we are taking some tentative steps in that direction. In 2013, the London Science Museum unveiled "REX" (robotic exoskeleton), a completely manufactured "cyborg" consisting of organs and organ systems from laboratories and companies around the world. These included artificial eyes and kidney (University of California, U.S.), ears (Macquarie University, Australia), trachea (Royal Free Hospital, U.K.), heart (SynCardia, U.S.), spleen (Yale, U.S.), pancreas (De Montfort University, U.K.), hands and arms (Touch Bionics, Scotland, and Johns Hopkins University, U.S.), blood (Sheffield University, U.K.), feet and ankles (Massachusetts Institute of Technology, U.S.). REX stands with the aid of a bilateral leg robotic exoskeleton.

Dan Ferris, professor of neuromechanics at the University of Florida, works at the interface of biomechanics, neuroscience, and human movement. Dan foresees a future in which "cyborgs will be more common than not. Artificial body parts, including bionic arms and legs, artificial eyes and ears, and even brain computer chips for boosting memory or other mental abilities will be accepted and readily available. The line between human and machine will be blurred. The ethics of medical practice will be more difficult as [the basic medical precept, to] 'first, do no harm' will be hard to interpret." It's conceivable that these technological advances could bring an end to physical disabilities.

Yet Ferris has concerns that inequities in our society may be worsened because "only the rich may have access to state-of-the-art technology to cure physical disabilities."

Thomas Stieglitz tends to agree. He is a leader in neurotechnology in the Department of Microsystems Engineering at the University of Freiburg in Germany and has been at the forefront of brain–machine interface research for many years. Much of his work has addressed the critical problem of how to produce and maintain effective connections between artificial implants and human biology. Stieglitz told me that the next century will bring "technical implants that interface nerves with minor foreign body reaction and allow selective recording of nerve signals and electrical stimulation of nerves. . . . Energy to power these devices will come directly from the body." This seamless interaction holds tremendous promise for targeted medical and rehabilitation therapies.

Stieglitz worries that this technology could be misused. Imagine, for example, someone using assistive technology meant for the treatment of disease to create human cyborgs for malign purposes. Stieglitz envisages "non-medical lifestyle or soldier enhancement, e.g. hearing colors or infrared vision."

Research along these lines is progressing at a stunning rate. In *Inventing Iron Man*, I made the argument that for the exoskeleton of Marvel's character to truly function as we see it in movies and in graphic novels, it would have to be a neuroprosthetic controlled by neural commands from the spinal cord and brain. That is, the ultimate brain–machine interface connecting a human to a powered robotic exoskeleton.

In framing a technological superhero in this way, I made a number of what I believed to be speculative observations. For instance, to work seamlessly, a human exoskeleton like that of Iron Man needed to have sensory feedback. The suit would have to function as a kind of synthetic sensory skin that would feed back into the sensory cortices via the brain–machine implant. I also observed that connecting a neurological implant in the brain to control an exoskeletal system would require bidirectional information flow. This is to say, if the controllers for the exoskeleton were "hacked" (as happens routinely in comic books and science fiction movies!), someone would be able to control the human user through the same interface.

I thought it would take years for these predictions to come true. Not so much. In 2011, before *Inventing Iron Man* was published by Johns Hopkins University Press, Miguel Nicolelis's group published a paper in the scientific journal *Nature* showing that learning with brain–machine interface control could be enhanced in a mouse when sensory stimulation was included in the design. The same group went on to demonstrate real-time sharing of behaviorally relevant sensory information between the brains of two rats located in separate geographical regions (Natal, Brazil, and Durham, U.S.). While this design didn't actually use the "hack" of remote-controlling one rat from a distance, it was proof of exactly this principle.

WELCOME SUPERHERO OR A BRIDGE TOO FAR?

At the start of this chapter, I quoted a line from *Amped* to the effect that change can occur imperceptibly, as slowly as mold develops on bread. In the same way, the slow invasion of implants and assistive devices is occurring almost without anyone noticing. Daniel H. Wilson's science fiction novel takes place in a dystopian future where neural implants (most notably something called the "Neural-Autofocus" used to sharpen concentration and intelligence) become widely available.

In the beginning of the novel, these devices are introduced

for use in those with cognitive disabilities, mental challenges, or health risks (for example, to control epilepsy), but eventually they see widespread application throughout the population. Two classes of humans emerge: those who are "amped" and those who are not. Wilson addresses many ethical and moral issues in this engaging novel. Soon we will have to address them in real life. The parts of human function that we are able to modify, restore, or enhance continue to expand. Figure 4 illustrates some head-to-toe examples of research applications and devices we will explore.

When they were first developed, most of these technologies were described in terms of restoration and rehabilitation. Some, like Google Glass, were initially launched as enhancement or entertainment tools but now are viewed with an eye to their medical applications. In any event, many if not all of these technologies could be used as functional enhancements that take us beyond the normal range indicated at the far right of the performance continuum in Figure 3.

Molly Shoichet is a tissue engineer and professor of chemical engineering and applied chemistry at the University of Toronto who studies polymers for drug delivery and regeneration to promote healing. She thinks that regenerative medicine will truly come of age in the next hundred years. "We will be able to fully replace and seamlessly integrate artificial body parts in our bodies. Diagnostic testing will help us identify, diagnose, and treat neurological disorders in utero, thus overcoming some of the most devastating childhood disorders like cerebral palsy."

There will be dramatic improvements in stem cell delivery, manipulation of the stem cells without our bodies, and inducing our normal mature cells to revert to pluripotency—the state where they could become many other cell types. These techniques will allow us to treat and effectively cure neurological disorders and other diseases (cardiac, liver, diabetes), and disorders associated with trauma and aging, such as stroke, spinal cord injury, and Alzheimer's or Parkinson's disease, thus improving quality of life for our species.

Shoichet worries that future advances in regenerative medicine will be misused by the rich, with the goal of attaining immortality. They could also be used (or misused) to make a stronger species—not to treat disease or disorders, but rather to make us stronger and smarter—manipulating biology to change the nature of our species.

THE GROWING PAINS OF LIONEL MESSI, FC BARCELONA MIDFIELDER

We have a high-profile, real-life example of the power of hormones in Lionel Messi, the football phenom with FC Barcelona. Messi is one of the best to ever play football (*soccer* in North America). He has scored an absurd number of goals, including more than 60 in 2012, and broke the club record for most goals. He has been named world player of the year multiple times. And it was all enabled by growth hormone injections.

According to Guillem Balague in his book *Messi* (2013), Lionel Messi was diagnosed with a growth hormone deficiency at age 11. Growth hormone comes from the anterior pituitary gland, situated at the base of the brain. The pituitary is about the size of a pea, but it is hugely important. The hormone it produces is a protein hormone that helps to regulate cell production. Low levels of growth hormone lead to weakness, poor energy levels, bone weakness, and the risk of cardiovascular disease.

An individual with growth hormone deficiency either won't reach their possible size or will get there very slowly. Because it can help stimulate growth and recovery, growth hormone is

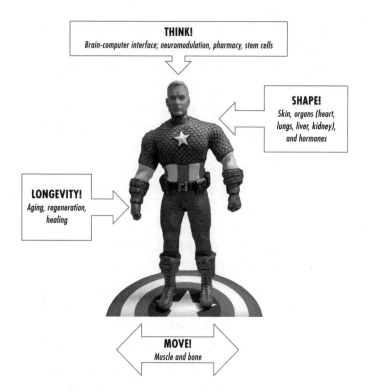

FIGURE 4: The super soldier procedure that created Captain America represents the accelerating evolution of human function.

considered an "ergogenic aid." It is on the list of banned substances published by the World Anti-Doping Agency and in most sports.

Growth hormone treatments allowed Messi to achieve his size in a more normal time frame than would otherwise have occurred. At 5'7", he's not exactly a giant in physical stature, but he has become a giant of global football. But what if you were growing fine, or even were fully grown, and took growth hormone anyway? What if Messi was actually 5'7" to start but wanted to play basketball in the NBA?

I often pose this question in public presentations. My audiences range widely in number, age, and background. Despite their heterogeneity, there is almost invariably an approximately 50/50 split in responses. That is to say, people are split when it comes to using growth hormone to enhance "normal" height, but no one objects to using it to remedy a pituitary gland deficiency. What's the ethical difference? Why do people make this distinction?

We are on the path to creating a real-life Captain America. We are rapidly moving toward the "superman" of Nietzsche, overcoming the limits of our species through the application of our swiftly increasing technological ability. We'll explore these realized and potential applications, and the questions they raise, in the chapters that follow.

Using the framework presented in Figure 4, the contents of the rest of this book are organized around the human capabilities we want to change, based in part on the traits associated with some comic book superheroes. We look at strategies to push aside the physiological limitations of our brains, muscles, bones, skin, and organs to enhance:

- how we think

- how long we will live

- our body shape and how we look

- how we move

We will explore how our species—"puny humans" in the words of the Hulk—is actually endangered by applications of science and engineering that take us well beyond natural functional ability and toward something truly superhuman.

Before we get to that, let's investigate the advances we've already made on the things we can see with our eyes—our body shape and how we look.

4. SHAPE!
CAN WE A.I.M. TO MAKE THE STAR-SPANGLED AVENGER?

Enhanced strength, speed, and endurance, as well as
proficiency in many combat disciplines and martial arts . . .
—Captain America described in *The Avengers: The
Ultimate Guide to Earth's Mightiest Heroes*

Captain America is not Thor, or Hercules, or the Incredible
Hulk—he is merely a man—honed to physical perfection . . .
—"The Queen, My Lord, Is Dead," in *The Secret Defenders* #7

In 1969, the Rolling Stones informed us of something we should
have known anyway: that we can't always get what we want. I
don't think Mick Jagger and Keith Richards were referring to the
form and function of the human body, but when we look at who
we are going to be, we start with what we got from our parents
and then work from there. And if Mick and Keith are right, we
might just get what we need.

Being unhappy with what we got from Mother Nature is a common human characteristic. Many of us are discontented with how we look. Or we wish our body were stronger, sleeker, or more toned. Generally, we have to make the best of it. We just have to accept that, within limitations, what's in our jeans is from our genes. Or . . . perhaps we could have surgery?

MAGIC MIRROR ON THE WALL, CAN WE MAKE THE FAIREST ONE OF ALL?

There is evidence that humans in India attempted to modify physical appearance through reconstructive surgery 25 centuries ago. Fiction writers got into the act before modern medicine really grasped the possibilities. And some carried the idea of body modification to extremes. When Mary Shelley penned her marvelous, paradigm-shifting novel *Frankenstein* in 1818, she gave vivid expression to the idea. Of course, this didn't work out very well for her creation, widely considered a monster.

It wasn't until the early 20th century that the idea of plastic surgery began to hit its stride. Walter Yeo lost both his eyelids during a World War I naval battle and surgery was performed to construct new "skin-flap" eyelids. Later, surgical techniques to replace lost tissue—functional surgery, we might call it—morphed into surgery strictly for cosmesis. That's the scientific way to say "for better looks."

Nowadays, cosmetic surgery is available to almost anyone willing to pay to have it done. Or done too often. The late American comedian Joan Rivers had her 734th cosmetic surgery when she was 78. These procedures can produce unwanted changes. For example, Rivers's last surgery was undertaken to treat voice changes that were a side effect of earlier procedures.

The American Society of Plastic Surgeons has a list of cosmetic procedures that it stresses are a personal choice. These discretionary surgeries do not affect function and are not meant to be undertaken to fulfill an ideal or the expectations of others. This list includes:

- "lifts" to reduce sagging arm muscles, breasts, eyelids, neck, face, abdomen, and thighs

- "implants" or "augmentations" to modify or increase the shape of the face, breasts, legs, and buttocks

- liposuction to reduce deposits of body fat in many areas of the body, including stomach, thighs, and arms

- rhinoplasty to modify the shape, size, and form of the nose

- modifications to the skin, including laser skin resurfacing, dermal fillers, and hair replacement

A few individuals see surgery as a way not just to become beautiful but—if not actually to become one—to at least look like a superhero. In Manila, Herbert Chavez had always imagined being Superman. Since 1995 he has had a host of cosmetic surgeries to reshape his chin and nose and enhance his thighs. Now he looks like Christopher Reeve, who was among the first modern actors to play Superman on the big screen. Chavez wears a Superman outfit at his costume store and tries to inspire other people to achieve superhero greatness.

THE REAL ARMS RACE

Surgeons have been developing techniques not just to rescue or change appearance, but also to alter form and function. On July 17, 1974, Los Angeles Dodgers pitcher Tommy John (288 career victories, ranks seventh all time among left-handed pitchers) tore the ligament in his left elbow. At the time, this injury typically spelled the end of a pitcher's career. Not in this case, though. That's because on September 25, 1974, Dr. Frank Jobe performed ulnar ligament reconstruction surgery for the first time. Jobe, an American orthopedic surgeon who had been an army medic in World War II, replaced the ligament in John's elbow with a tendon from his right forearm. Observers reckoned at the time that this experimental surgery had roughly a 1 percent chance of a successful outcome. He beat the odds: Tommy John returned to the pitching mound in 1976, after more than a year of rehabilitation. He pitched for 14 more seasons and accumulated 164 of his 288 career victories before retiring.

ESPN reports that Tommy John surgery now is estimated to be effective in as many as 85 percent of cases. According to *USA Today*, in 2003 almost 700 major league pitchers had undergone the procedure in the preceding two seasons. Sometimes players experience a 5–10-degree loss of range of movement post-surgery (which means they cannot fully straighten the arm), but this apparently does not significantly hinder throwing function.

It's an amazing example of useful functional surgery. But also important is the way the operation has been perceived. As C.S. Ahmad and colleagues reported in the medical journal *The Physician and Sportsmedicine* in 2012, some believe that the surgery can be used not only to restore function, but also to enhance pitching performance. It's possible that the perceived improvement is due to intensive post-op rehabilitation, but there is no question that some baseball insiders credit the operation for making good players even better.

GREAT SCOTT! THERE ARE HORMONES EVERYWHERE!

"Water, water, every where / nor any drop to drink." So said the ancient mariner in Samuel Taylor Coleridge's famous poem. (By the way, "Rime of the Ancient Mariner" is also the name of an excellent, epic 13-minute Iron Maiden song from their 1984 album, *Powerslave*.) How is this relevant to the present discussion, you ask? Well, your body is made up of a whole lot of water: each of your 100 trillion cells is 65–90 percent H_2O. The water in your cells, the internal environment in your body, and the water in blood, your plasma, can make up 50–75 percent of your body mass. A powerful way to change the function of the human body is to change the internal environment. That brings us to hormones, blood-borne messengers, and issues related to stress and steroids, all of which would need to be altered in any procedure designed to produce a super soldier like Cap.

Human physiological function is based on interactions in many complex systems and subsystems. But each responds to environmental stresses in a useful, and generally predictable, way. Strength is a good example. The expression of your muscle strength is a complex response to the stresses you apply to your body and the ability of your body to respond. The stresses might, for example, take the form of loads you pick up and carry. The response is to make more muscle tissue so that the mechanical stress on the muscle is effectively reduced.

Living organisms—from black widow spiders to Black Widow Avengers—respond to physiological stress in a way that minimizes the effect. The body seeks to maintain internal balance and equilibrium (homeostasis). If the stress is too large or too frequent—if, for instance, an athlete overtrains—their body won't be able to adapt and there will be a steady (and sometimes rapid) decline.

The failure to maintain homeostasis is what some athletes experience when they say they are "burnt out" in the period leading up to major events like the Olympics. Ideally they feel burnt out after their events, not before, but it doesn't always work out like that. Mistakes in the design of a training program sometimes have this unfortunate result. When challenges to homeostasis cannot be met, or if they are sustained for too long, biological organisms decline. They can even die.

The ability of each person to respond to exercise stresses is determined in part by hormones. Testosterone, cortisol, growth hormone, insulin-like growth factors, and the catecholamines are very important for regulating the muscle response to exercise and injury.

Doping and the desire to keep competition "clean" have become a thorny issue in both professional and amateur sports. *Clean* in this context tends to mean that the athletes have not resorted to using supplements like steroids that are known to affect the pathways that are helpful for their sports, whether the emphasis is on strength, speed, or endurance.

The definition of *clean* and how vigorously related policies are enforced varies. When Bud Selig, commissioner of Major League Baseball, appeared before the U.S. Congress in 2008, he supported testing baseball players for human growth hormone "when a valid and practical test becomes available." A valid test was in place and had been widely used in the Olympics when Selig testified, but it wasn't until the summer of 2010 that testing of baseball players began. Minor league ballplayers now are subjected to random tests for synthetic human growth hormone—but major league players still are not.

Cortisol is thought of as the "stress hormone" and is important in regulating blood sugar levels, blood pressure, and activity of the immune system. It is sometimes referred to as a catabolic hormone: over the long term, increased levels of cortisol in the blood lead to muscle breakdown. A side effect of taking anabolic steroids is that they interfere with cortisol and reduce its release into the blood. Cortisol levels are tightly controlled in the body, and levels in trained and untrained people aren't much different—unless the individual is over-trained and unwell.

How the human body responds to an adaptive stimulus, such as strength training, is determined by a number of complex variables. The response may be altered by changing

- the stress (the exercise)

- the level of hormones helping to provide the adaptation

- the level of in-between products for all the hormone pathways

- the regulating mechanisms in the body that alter these levels

This is not an easy issue to sort out or police. Your body is like a factory in which products such as muscle are made with the help of hormones at various points in the physiological process. There was a furor in Major League Baseball about androstenedione, or "andro," that largely centered on Mark McGwire and his chase to break the Roger Maris home run record. Andro is a precursor used in the metabolic pathway to produce testosterone. Using andro isn't "taking steroids," but the increased availability of androstenedione in the body can boost the production of testosterone. So, using the factory metaphor, you can create more muscle protein if you have enhanced levels of andro. Which is why it became such a big issue when it was revealed that McGwire was taking andro during his big home run push. McGwire was accused of doping. But androstenedione was not a banned substance in the league at the time and it isn't technically a steroid. It was easily purchased as a legal supplement.

Did andro affect McGwire's performance? Probably. It would certainly have helped with recovery. But, as pointed out by Howard Bryant in his book *Juicing the Game: Drugs, Power, and the Fight for the Soul of Major League Baseball*, the reaction was "emblematic of a sport in need of a strategy." Major League Baseball has yet to take a coherent and scientifically grounded position on doping.

We tend to think of doping as a modern issue, and that athletes and their fans have somehow become more ethical. But as sportscaster Joe Buck wrote in his 2016 memoir, *Lucky Bastard: My Life, My Dad, and the Things I'm Not Allowed to Say on TV*, "I don't think that the guys in the sixties and fifties were better human beings and wouldn't have taken it. They just never had the opportunity to take performance enhancing drugs." Our societal reaction to "doping" (both real and implied) seems to be more about where we draw a (fuzzy) line separating acceptable from unacceptable. At one extreme, having a poor diet lacking in protein and the required amino acids for muscle synthesis would lead to poor muscle growth. At the other extreme, we have the concept of doping, where anabolic steroids can artificially enhance muscle protein creation. It's complicated when you think about it. After all, while there are rules covering the use of supplementation in sport, there are no such rules in life.

WHAT DOES IT MEAN WHEN WE SAY SOMETHING IS "IN OUR GENES"?

As a sub-branch of biology, genetics is all about studying and explaining the physical traits or attributes passed from parents to their kids. The things we can actually see in a person, such as Steve Rogers's blue eyes, are called phenotypes. These are the physical expression of our genes found in each of our genotypes—our own unique genome. Although words like *genome*, *proteome*, and *metabolome* sound modern, the concept of inheritance of physical characteristics—those phenotypes—has been around for a long time.

About twelve thousand years ago, the Egyptians bred and domesticated dogs, sheep, goats, and camels. As part of this breeding process, inherited traits were passed from parents to offspring and were easy to observe in the physical form and behavior of the animals.

Fast forwarding a bit, although it wasn't yet called genetics, most of the work that led to our present knowledge of genes was carried out by one person. This labor-intensive and time-consuming process was undertaken by a quite unassuming individual who became the first rock star of genetics. Okay, he wasn't exactly a rock star—he was actually a monk—but back in the mid-19th century Gregor Mendel absolutely did lay the foundation for modern genetics. He liked to experiment with plants, particularly beans and peas. His extensive studies based on his observations of more than ten thousand plants over almost a decade were published in 1866.

If Mendel can rightly be thought of as the father of modern genetics, he can also be thought of as the father of systematic genetic engineering. That is, he attempted to create, in a very deliberate way, targeted changes in a living organism by breeding for desired characteristics. What Mendel didn't understand was the mechanism for inheritance—how it actually worked.

Although Mendel didn't know it, there were others doing research that would ultimately be relevant to his findings. A Swiss physician, Friedrich Miescher, discovered in 1871 that genetic information is stored inside the nuclei of our cells as DNA (deoxyribonucleic acid). The fact that genes are completely composed of DNA had to wait for Oswald Avery's discovery in 1946. DNA represents the physical embodiment of the genetic code. To fully understand how it worked and to think about using the information, scientists needed to understand the structure of DNA itself. Think about human anatomy. If you don't know what bone's connected to what, you can't really figure out how things work. You need the structure first.

ROUNDING ALL THE BASES ON THE WAY HOME

The official scientific documentation of DNA structure came in 1953, when James Watson, Francis Crick, Maurice Wilkins, and Rosalind Franklin published groundbreaking papers on the double helix and the implications of the pairing of nucleotides (also called base pairing). One of the most famously understated sentences in the history of science is found at the end of their report: "It has not escaped our notice that the specific pairing we have postulated immediately suggests a possible copying mechanism for the genetic material." That phrase was an English way of saying, "Umm . . . we think we have discovered the biological building blocks of life as we know it." Watson, Crick, and Wilkins won the 1962 Nobel Prize, and their work became the foundation of modern genetics.

Amazingly, given the diversity we can see and experience in human form all around us, there are only four bases found

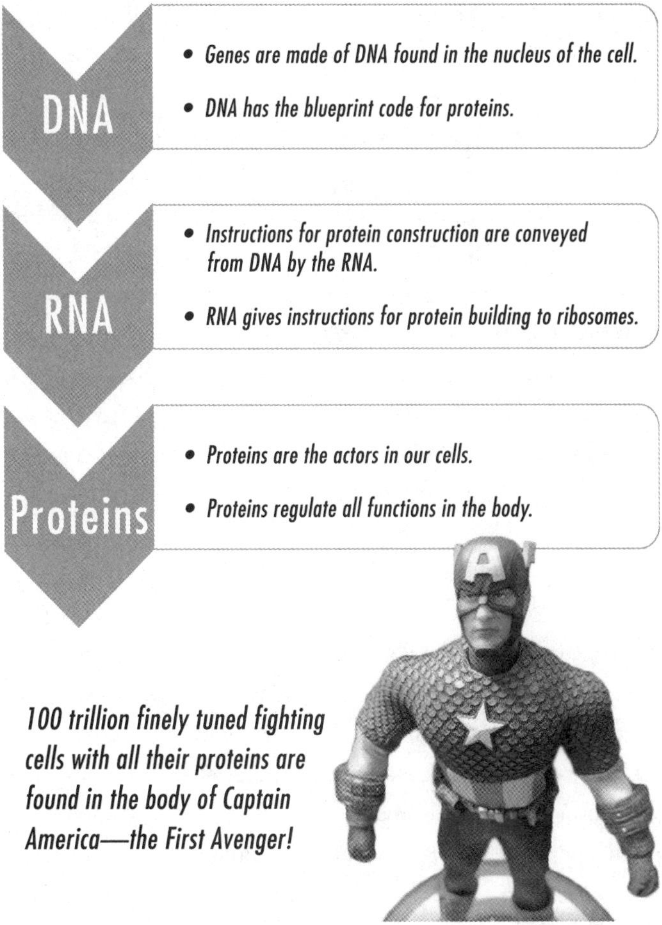

DNA
- Genes are made of DNA found in the nucleus of the cell.
- DNA has the blueprint code for proteins.

RNA
- Instructions for protein construction are conveyed from DNA by the RNA.
- RNA gives instructions for protein building to ribosomes.

Proteins
- Proteins are the actors in our cells.
- Proteins regulate all functions in the body.

100 trillion finely tuned fighting cells with all their proteins are found in the body of Captain America—the First Avenger!

FIGURE 5: The power of proteins unleashed by the coding within DNA.

on a single strand of DNA. These base nucleotides are adenine (A), thymine (T), guanine (G), and cytosine (C), and they bond only as A-T and C-G. We hear a lot about DNA and genes these days. The physical bit that we think of as genes is formed from those tiny nucleotides that are put together to form chromosomes. In your body you have around three billion pairs of these nucleotides. All together you have between 25,000 and 50,000 genes all arranged on 23 pairs of chromosomes. The job of your DNA is to instruct the building of proteins.

It is a very efficient packing job. Each strand of your DNA is about six feet long, but they're stuffed into nuclei of cells you can't even see! This occurs with the help of proteins called histones. They are the "Neat Nellies" of biology and help keep your genetic material as organized as possible. The collection of those genes in your chromosomes allow you to be the person you are right now, with all the physical attributes you possess. These genes can be slightly altered to take on different forms, called alleles. The sets of alleles you have in your genes are what give you your specific genotype.

A LITTLE MORE ABOUT ALLELES

Your alleles are either dominant or recessive, but all together they are who you are. Usually there are more than two alleles involved, but if you have at least one dominant allele for a certain trait—for example, one affecting muscle strength, like myostatin—that trait will be expressed. To say that an allele is dominant is like saying the major function of that gene is activated. You only get to "see" a recessive trait if you have recessive alleles in both pairs. Keep in mind—and the very important implications of this will keep coming up—that your genes (your genotype) and the physical expression of those genes (your phenotype) are not the same thing. All of this sets the background for thinking about your genotype as regulating what you can become, but not necessarily what you must be.

Until recently, there was nothing you could do to change how your phenotype would be expressed. Now, in the burgeoning era of genetic engineering, we are moving closer and closer to being able to alter the genetic coding in order to change the proteins produced.

THE POWER OF PROTEINS

The function of DNA is to give instructions to the factories in your cells, the ribosomes, so they can make proteins. Proteins are crucial to life. They work in your body to shuttle things within cells, to work as enzymes, to provide structural support, and as biological motors. You might think of them as biological words. When making a protein (something that is constantly occurring), you use an "alphabet" of 20 amino acid "letters" that can be combined in many different ways. Many of your protein words are really long—way longer than, say, "supercalifragilis-ticexpialidocious." Some have hundreds of amino acid "letters."

A metaphor for protein production is computers and printers. You can have the plans for a biological organism on your hard drive, but you have to print it before it can be acted on. The DNA inside Steve Rogers is his hard drive, and the image projected on the screen of his computer is messenger RNA or ribonucleic acid (explained below). If he wants to make something, he hits Print. The printout is a polypeptide and the printer itself is the protein-making parts of the cell—the ribosomes. The paper is the actual protein you are trying to make. This metaphor falls apart a bit for Steve because he doesn't use a computer. He does all his communicating with a big red shield.

The punctuation (to pick up the grammatical metaphor again) comes from the four bases (A,T,C, and G) and two base pairs (A-T and C-G). Three base pairs together give us a codon, and the codon indicates where to start and stop the making of a protein. This requires a cousin of DNA, called ribonucleic acid, or RNA. RNA is constructed like DNA and also has four bases. For RNA the thymine (T) is replaced by uracil (U).

To do all the copying from the DNA to the place where proteins are properly made, you need three kinds of RNA for transcribing and translating. The DNA is kind of like the blueprint kept in a library that you can't check out and take with you. You have to make elaborate copies and you do that in your cells using RNA. First up is RNA for transcribing the DNA coding. During this process the DNA strand is uncoiled so the gene segments are revealed. Transcriptional RNA then forms the complementary and proper base pairs that match those of the DNA.

Messenger RNA binds to the protein-building ribosomes and, with help from transfer RNA, we get a new protein. This is occurring constantly in the cells of all living organisms. This gives us the complex life that we see, hear, and feel all around us—and that we now have more and more ability to control.

SUPERHERO MOLECULES OF LIFE

This brings us to the issue of what all those letters and words look like and how much of an alphabet Steve's parents, Joseph and Sarah, gave him. Usually the tongue-in-cheek expression of "choose your parents wisely" comes up here. Since the success of the human genome project, we now have a lot of information about the role played by specific genes in affecting athletic and exercise performance.

The human genome was mapped in 2005 and we now have complete genome maps for many different species. The Human Genome Project mapped slightly more than 90 percent of the human genetic code and was accomplished through both publicly funded and private commercial efforts. Genetic samples from fewer than 12 people were used in the enterprise, and the public project samples were taken only from one person (code-named RP11) from Buffalo, New York. Work continues to flesh out the full genome and its variations. This is crucial for what we are focusing on—that is, exceeding the "limits" of the genome—as it has implications for drug treatment, response to the environment, and the extent to which we can change who we are—that is, nature versus nurture, which also includes technological intervention.

Gene expression for a given trait involves mutations in the gene pairs. Changes in one allele, as discussed above, may lead to expression of a certain phenotype. It is important to realize that this means the gene may, but not necessarily will, be expressed.

This leads to a significant issue: the relative importance of genetics as opposed to environmental conditions in determining expression of phenotype. Or, put another way: how much of what is coded in our DNA defines our behavior, and how much of our behavior is shaped by what we experience? Humans can adapt to a range of conditions, but many factors beyond genetics have an important effect. There is variation, then, in the responses and adaptations that a person might have, and the genetic influences may not be the dominant factor. That means that the genetic effects will only be revealed, expressed, or maximized in certain specific situations.

CAN WE BE PROGRAMMED FOR DIFFERENT SHAPES AND LOOKS?

There has been an explosion of interest in genetics in recent years. The media remains full of reports about genetic engineering, gene therapies, prenatal screening, and the implications of the genome project for human life and society. For all that, however, our ability to change our genes has been limited.

Everything changed, however, in 2010, when Craig Venter and his team inserted synthetic DNA into a bacterial cell and watched it grow. This approach could be used one day to specify new human characteristics and traits, and could give rise to endless possibilities for changing what it means to be human—*including how we look*. The practical problem with trying to genetically engineer something like physical beauty is that there isn't a specific "beauty gene" that maps onto some aspect of a phenotype we'd all call beautiful. Beauty is more subjective than that. But scientists have used gene editing to change the appearance of certain other animals. In the process, they created "dino-chickens."

It's been more than 150 million years since some dinosaurs began to develop the characteristics of birds. Scientists who work in this subject area have focused their studies on the emergence of feathers and flight and, in some instances, intriguingly, on the evolution of the dinosaur's snout. In a study published in the journal *Evolution* in 2015, Bhart-Anjan Bhullar at Northwestern University, Arhat Abzhanou at Harvard, and their colleagues looked at a number of modern animal species whose snout or beak appeared to represent a link to the comparable bone structure of the dinosaur. Their sampling included alligators, emus, turtles, and lizards.

They paid particular attention to two genes that are linked to the development of the middle portion of the face of birds. These genes acted differently in birds than in reptiles. The scientists thought that if they could block the activity of these genes and the proteins they produced (called FGF and Wnt) in chicken embryos, they would see what the true snout of a bird—and thus their dinosaur ancestors—might look like. Their experiment worked very well. While the "dino-chicks" were not allowed to hatch, the embryos produced various versions of "snouts" that share similarities with dinosaurs such as Velociraptor and

Archaeopteryx. There was variation in the appearance of the embryos, however, so the effects were subtle.

This work was narrowly directed but has broad implications. It's still early days—the group was not actually able to alter gene function but only the protein expression. Nevertheless, this was a proof of principle. The researchers demonstrated that it will likely be possible to modify the outward form of animals sometime in the future. Which is relevant, of course, to all of us, because people are animals. And like all animals, we like to move.

Captain America is always moving—whether he's hauling away a Hydra agent or slinging his shield. So let's consider how we could improve his ability to move by amplifying the stuff that makes movement happen—his muscles.

5. MUSCLES IN MOTION!
STEM CELLS, STEROIDS, AND
THE SENTINEL OF LIBERTY

Hoped-for stem cell cures and the researchers who will
create them are like comic book heroes who use their
unique supernatural powers to rid the world of evil.
—Lawrence Burns, "'You Are Our Only Hope': Trading
Metaphorical 'Magic Bullets' for Stem Cell 'Superheroes'"

It was a signal moment not merely for sports
. . . but for human evolution.
—Michael Farber, on Ben Johnson's 9.79-second
100 meters at the 1988 Seoul Olympics

The super soldier procedure did its job well. Captain America
is physically imposing. He is big and muscular. As Dr. Erskine
says in Marvel's *Captain America: The First Avenger*, the procedure involved "a series of microinjections into the subject's major
muscle groups . . . the serum infusion will cause immediate cellular

changes . . . and then, to stimulate growth, the subject will be saturated with Vita-Rays."

Muscle has a lot to do with many of Captain America's adventures and stories. As a World War II POW trapped somewhere in the South Pacific, Cap actually outmuscled a giant ape in "The Vultures of Violent Death" (*Captain America Comics* #28, July 1943). Dr. Erskine's Super Soldier Serum had miraculous power. And now it has a contemporary analogue. Enter the stem cell, superhero of science. Possibly. Not since *Daily Bugle* editor J. Jonah Jameson began besmirching Spider-Man's character in *The Amazing Spider-Man* #1 (March 1963) has so much false hope, hype, and paranoia been written about a collection of human cells.

Talk of stem cells, stem cell research, stem cell therapies, and the ethics of stem cells is seemingly everywhere. Stem cells have been pitched as a medical breakthrough with an infinite array of potential applications. The Ontario Science Centre even ran an exhibit in 2011 called "Super Cells: The Wonder of Stem Cells," which included comic book–style graphics and images. But what can stem cells really do? The best way to start with the story is in muscle, where they were first identified.

STEM CELL SCIENCE RESERVE

There's a link between the Avengers—and many other superheroes—and our attitude toward stem cells and stem cell research. Lawrence Burns (see the epigraph to this chapter) captures part of what I mean in "'You Are Our Only Hope': Trading Metaphorical 'Magic Bullets' for Stem Cell 'Superheroes.'" Just as those who look to "Earth's Mightiest Heroes" to save them from imminent peril, so do those who look to science to save them from a range of illnesses see stem cells as their only hope.

The 2012 *Avengers* movie included a re-envisioning of the origin story that's a lot closer to the more recent (and great) Marvel Ultimates story lines. The cast luckily includes some of the original founding members—Iron Man, Thor, the Hulk, and Captain America—but also Black Widow and Hawkeye, who were added later in the comic books.

What's always been really compelling about the Avengers is the spectrum of superheroes included in the team. You have on one hand a hero like Iron Man, who represents a more "realistic" superhero. As I wrote in *Inventing Iron Man*, his origin story has some very plausible bits to it. On the other hand, we have Thor, the Norse god of thunder. Since he's a mythological figure, this makes him one of the least relatable, least realistic characters. The way Thor is pitched in recent imaginings (including the excellent Thor movies) is inspired. I especially like the way the screenwriters have worked in the concept contained in the comment by science fiction legend Arthur C. Clarke, that "any sufficiently advanced technology is indistinguishable from magic," in order to bridge Midgard (earth) to Asgaard.

Midway between Iron Man and Thor on this continuum, we have Captain America. Cap is basically like DC Comics' Batman in terms of fighting prowess and overall skill. Captain America would have had to undergo a training regimen similar to the one I outlined for Bruce Wayne in *Becoming Batman*. (Except in the case of Cap you add in stem cells, steroids, and gene deletions. This is the focus of Chapter 8.)

Superheroes like the Avengers seem to have unlimited power. They transcend humanity with skills beyond our reach and save us from peril—just as stem cells may potentially be capable of healing just about any disorder or disease. Stem cells, like the Avengers, are also seen as our only and final recourse for getting the help we need. Saving the day in the comics and saving lives in real life.

Just like stem cells, the Avengers are held in reserve until they are needed. There are some interesting parallels in the opposition and controversy that stem cell research and the Avengers have faced. In order for the Avengers to save the earth from evil (or sometimes their own members—yes, we're looking at you, Hulk), they need to be allowed to get in there and do the job. Too often, they are opposed by a public duped by PR factions and misplaced religious concerns that pitch them as going above the law. Unfounded opposition has held back progress. For example, concerns about how stem cells are derived and moratoriums on certain cell lineages blocks research and delays discoveries of effective treatments for human diseases and health problems.

According to Lawrence Burns, "those who stand in the way of the superhero are unwittingly aiding the cause of evil." This is a time when the often misapplied saying, "if you're not with us, you're against us," actually applies.

SUPERHERO SILLY PUTTY

Silly Putty wasn't designed to be the fun silicone polymer that many have come to know and love. In fact, it wasn't actually designed for any of the applications it is used for now. Although there's some controversy about the original patent and who should get the credit for it, Silly Putty has been around since World War II. It was an accidental by-product of experiments aimed at creating a substitute for rubber to be used in the war effort. As it turned out, Silly Putty has countless useful applications, including as a rehabilitation tool to improve dexterity after stroke and for helping Apollo astronauts in 1968 keep their tools from floating around in zero gravity.

Like Silly Putty, stem cells are now viewed as a panacea for just about everything, from Crohn's disease, baldness, wounds, and hearing loss to spinal cord injury and stroke. There is currently a major initiative to use stem cells to regrow heart muscle tissue. Articles with catchy titles like "Can Stem Cells Mend a Broken Heart?" abound.

What is now a stem cell industry began with a discovery made by Canadian scientists Jim Till, Ernest McCulloch, and colleagues. Their 1963 paper in *Nature*, obscurely titled "Cytological Demonstration of the Clonal Nature of Spleen Colonies Derived from Transplanted Mouse Marrow Cells," had a spectacular—both positive and negative—impact. On one hand, stem cells have no real peer when it comes to the promise they seem to hold. On the other hand, controversy related to their study and use has been intense from the start.

Stem cells are cells that can, with the appropriate cues, develop into other cell types. They can function as an excellent internal resource for cellular repair. Unlike other "differentiated" cells that you already have in your muscle, brain, and bone, stem cells can still divide and reproduce to produce new cells. But how do you go from stem cells in a laboratory dish to useful application inside a human? The first step is to get hold of some—a donor is needed. One way to achieve this is through a bone marrow transplant: a donor gives some of the stem cells in their bone marrow for the benefit of someone else. The donated cells then get purified. This part is all fairly straightforward, but the next steps aren't.

It's important to get a large population of cells, so they need to be expanded to whatever size is needed to replace the organ you have in mind. Then they need to be delivered to the right place in the body, and then made to grow, or "home," in that organ. These last steps are the hard ones. Growing the population to the right size and then taking them out of the medium is tricky, and getting the cells to home is also difficult. It's kind of like trying to transplant the most unstable plant you can imagine. It's hard to grow in the first place, and then transplanting it to a new pot is often fatal.

Consider, for example, stem cell applications in muscle tissue—which is fitting because stem cells, originally called satellite cells, were first discovered in muscle tissue. They were called satellite cells because they were thought to functionally orbit the cells in muscle.

Work using stem cells to become specific—programmed—cell types is challenging. It's hard, for example, getting them to become the cell type you need—such as a hepatocyte in the liver. You then have to get them to form something useful—such as an actual liver. Scientists all over the world are working to figure this out. In 2011, Mototsugu Eiraku and colleagues at the RIKEN Center for Developmental Biology in Kobe, Japan, published a paper in *Nature* entitled "Self-Organizing Optic-Cup Morphogenesis in Three-Dimensional Culture." Hidden in that title was an amazing advance. Yoshiki Sasai with the RIKEN group worked for more than 20 years on a special project. He wanted to get embryonic stem cells from the mouse to form organs—all by themselves. He tried first with brain tissue, but more recently he has been working on the perfect "mix" (some have actually called it a "recipe") to get stem cells to form eyes. His recipe included the stem cells themselves, a kind of "protein mix" called laminin, and growth factors. Initially he wasn't surprised that the cells clumped together to form round shapes that became hollowed-out spheres. This is a common growth outcome of stem cells in the dish. What was uncommon was that growth continued to diversify and become more sophisticated.

After seven and nine days, the spheres began to look more like goblets, and two cell layers formed. Just as in the real eye, the cells developed a pigmented layer and a layer of retinal neurons. He created "optic cups," which are the developmental precursors to the retina. Sasai had finally succeeded in his special project. Work continues to apply this technique across species and with different target organs.

There are some big challenges to usefully applying stem cell therapies. These challenges remain whether we are thinking about using stem cells for treating disease or for altering and enhancing healthy humans. Paolo Di Nardo, Dinender Singla, and Ren-Ke Li, stem cell scientists from Italy, the United States, and Canada, recently summarized the main concerns in the *Canadian Journal of Physiology and Pharmacology*. They point out, for example, that it can be difficult to control cell growth and prevent tumors.

It would be wonderful if stem cells could be used to replace defective organs, such as the heart, and a number of scientists have investigated this possibility. One idea was to use the connective tissue of the heart as a frame on which to grow new heart muscle cells. But it's been a bit more difficult than anticipated. Once stem cells have been coerced to form a certain cell type, it can be a challenge to achieve a balance between, on the one hand, getting stem cells to form the desired tissue—in order to replace an organ—and, on the other hand, keeping them from growing too much and becoming a tumor. In work with heart tissue, the problem has been to grow the desired tissue. Only about 10 percent of stem cells put into the myocardium actually survive and stay there. It's a bit like building a sand castle: it may start big and be well defined, but water and gravity erode it fairly quickly.

MYOSTATIN AND THE TRUTH BEHIND SUPER SOLDIER STRENGTH

As we know, Cap starts out as 90-pound weakling and is transformed into a super soldier. What if—in real life—you could tap into and unleash some inner mechanism, the equivalent of Super Soldier Serum, inside your muscles? What if it could be like removing the shackles from Hercules and allowing for dramatically increased strength? Hey! Cap, like Hercules, even has super-strong neck muscles! They saved him from sure death by strangulation back in "The Spawn of the Witch Queen" in November 1942 (*Captain America Comics* #20).

Which brings us to two proteins that go by the could-be-created-by-Stan-Lee-sounding names of myostatin and activin A. Myostatin and activin A are found in your muscles and they are both chalones—factors secreted by your cells to suppress excessive growth. They function to keep the size and number of your muscle cells—and thus your overall strength—within a certain "human" range. Because their presence holds growth in check, removal of these factors allows muscle cells to get larger and increase in number. This is where super strength could come from.

Although it wasn't known as myostatin back in 1807, the effect of this growth factor was initially described in cattle, as "bovine muscular hypertrophy," by a British farmer, H. Culley. Cattle that have a myostatin gene deletion look unusually and excessively muscular. They're sometimes described as "double muscled." A double-muscled cow has less bone and fat and much more muscle than its bovine "single-muscled" neighbors.

Mutations of myostatin and activin A are found in many mammals, including rodents, dogs, pigs, and sheep. They're uncommon in humans. In 2004, the *New England Journal of Medicine* published a case study by Markus Schuelke and collaborators on the physical and genetic characteristics of a child with this mutation. Ultrasound images of the child show his quadriceps muscles are much larger than in the "control" boy. The authors reported that he continued to develop normally but had much greater strength than usual. When he was four and half years of age he could extend his arms straight out to his sides while holding a three-kilogram dumbbell in each hand.

So we know what happens in cases where myostatin and activin A are randomly deleted. What if we could somehow control this ourselves?

WHAT DO WE WANT TO SUPERCHARGE?

Myostatin is interesting because it works in the opposite way to what you would think. Myostatin controls muscle growth, but it does so by stopping skeletal muscle from getting too big. So myostatin normally inhibits the growth of skeletal muscle and acts as a kind of brake on muscle growth and strength gains. If this gene were "deleted" from Steve Rogers during the super soldier procedure, he'd have enhanced muscle growth and increased strength.

This brings us to the subject of super-strong monkeys—which actually scare me a little bit. That's because of this thing I've had about monkeys ever since I first saw *The Wizard of Oz* when I was five years old. It was the scene where the Wicked Witch of the West calls out her winged monkeys. Those creepy simians defeat the Winkies and the Great Oz and kidnap Dorothy. When I saw them come spilling over the rooftops and pick up Dorothy, I thought then and I still think now—monkeys should not fly. This pop culture reference is one of the only ones that Steve Rogers, shown in our time period, actually follows and "gets" when used by Nick Fury in Marvel's *Avengers* movie.

Janaiah Kota and other research scientists at the Center for Gene Therapy at Nationwide Children's Hospital in Ohio and at Ohio State University weren't thinking of flight when they carried out their work in 2009. They were focusing on strength.

They used a virus (a viral vector) to insert the human gene for follistatin into the knee extensor muscles of macaque monkeys. Follistatin works to block the action of myostatin. In other words, it removes the brake on muscle growth. This gene insertion permanently modified the muscle properties of the monkeys. Their muscle grew about 25 percent larger and stronger than normal. This kind of experiment had been done before in other animals, including mice, but this was the first time it had been successfully shown in nonhuman primates. While there are numerous issues to sort out—importantly, the need for immunosuppression to keep the injected animal's immune system from attacking the "foreign" inserted gene—this approach has huge potential to improve muscle mass and function in degenerative muscle disorders. It also provides a means to circumvent natural genetic regulation of muscle strength in us normal folks.

So a procedure for super soldier treatment could include either the myostatin gene deletion or the follistatin gene insertion. Or both. Continued advances in molecular biology have decoded many genes associated with high performance in humans, giving a kind of guide for targets we might want to mutate. As with many approaches we've discussed so far, the long-term effects of deliberately inducing genetic manipulations on growth factors, such as follistatin, aren't well known. Many of the approaches remain to be tested clinically in humans, but we are close to bioengineering super-strong monkeys. Which, as I said earlier, kind of freaks me out.

SUPER SERUMS AND THE SENTINEL OF LIBERTY

Levels of testosterone circulating in the blood rise during and after exercise and after a period of hard work. Testosterone also can amplify the effects of other hormones, like growth hormone.

So testosterone is kind of like the team leader of a superhero hormone group for growing strong bodies.

The synthesis and release of testosterone involve a number of steps and many "in-between" products that can stimulate the body to produce even more testosterone. All the systems are linked and can potentiate one another. This complexity partly explains the difficulty of testing for steroid abuse. The use of steroids and stimulants in unsanctioned contexts continues. Pat Mendes, for example, billed as the "world's strongest teenager" when he was just 19, bench-pressed more than 500 pounds. Just before the 2012 Olympic Games in London, this American weightlifter, once expected to win a medal at those games, instead tested positive for excessively high levels of human growth hormone.

Often overlooked in assessing the influence of substances like steroids is the effect they can have after they are withdrawn. Ingrid Egner and her collaborators in Oslo, Norway, examined this very issue in 2013. Their investigation requires a bit of preliminary explanation.

CELLULAR MEMORIES AND EPIGENETIC EFFECTS IN AN AVENGER

It has long been known in strength-training circles that someone who has trained extensively, and then stopped, can reacquire muscle mass and strength more rapidly than someone who hadn't trained at all. This was thought to be due to rapid changes in the nervous system affecting the coordination and activation of the muscles. This, in turn, might be related to "epigenetics" (recall the Hongerwinter during WWII from Chapter 2).

For the longest time, media reports focused exclusively on genetics and the human genome. Then we heard that the really important thing about the genome is how genes code for proteins—the building blocks of living organisms. That introduced us to the proteome. Next up was the idea that we need to understand what those proteins may do in life, which has a lot to do with metabolism. All the chemical processes going on in your cells as part of metabolism to keep you alive involve lots of enzymes (proteins) and by-products. Measuring all those small molecules and evaluating what they are is the metabolome.

The latest hot topic is the epigenome—the consideration of changes to the genome that aren't due to genetic influences. The epigenome determines which genes actually get activated and expressed by which kind of cells and when. If the genome contains the essence of your genetic potential, epigenetics is the way your potential is brought forward and used. The genome is like your dictionary full of words that aren't all used at once—and some never used at all. Epigenetics involves the process of pulling those words out and usefully applying them in sentences for conversations you need to have. Which brings us to the influence of environment. Epigenetics essentially bridges the gap between nature and nurture.

The term epigenetics describes how gene expression is regulated and what genes are expressed in an organism. This does not change the actual nucleotide sequences—the building blocks—in the genes. The epigenetic changes that happen to you in the course of your life affect the next generation of cells that you produce. Any epigenetic change may affect the next generation, but it doesn't necessarily have to.

While every cell in your body carries your genome, your epigenome has a number of flavors, depending upon the cell and tissue type. The key things about epigenetics are that it affects gene expression, changes during development (when stem cells are differentiating into the cells they are going to become), and changes in disease states.

Cancer, of all diseases, has been the one linked most clearly with epigenetic changes. For example, a gene that when activated produces unchecked growth of cells in the lungs—resulting in lung cancer—may only be activated and expressed when an environmental cue is present, like cigarette smoke. But biology isn't typically that simple and linking diseases directly to DNA changes is difficult. This is largely because changes that yield disorders often occur outside the parts of the DNA that code for proteins and that we understand better.

Now, remember Ingrid Egner and her colleagues in Norway? They were interested in the effect of training on muscle growth. Unlike most other cell types, those in muscle—the muscle fibers—have multiple nuclei. When strength training occurs, muscle mass increases and the number of nuclei in each cell also goes up. Egner's team wanted to know if this "cellular memory mechanism" could be influenced by steroids. They gave mice a testosterone derivative for 14 days, which produced about a 66 percent increase in nuclei and a 77 percent increase in the size of the muscle fibers. Three weeks after withdrawing testosterone, the size of the muscle fibers had reverted to the level found in animals that had never trained or been given drugs.

This part of the experiment was to cause a change in the "memory" within the nuclei of the muscle fibers. While the size of the fibers fluctuated, the number of nuclei remained elevated for three months after the testosterone was withdrawn. Did this mean that the muscle fibers would respond differently to training? That is, would they respond like "normal" or enhanced muscle because of the larger number of nuclei?

In the next part of the experiment, two groups of animals (the ones who were previously exposed to the testosterone derivative and the "control" animals that weren't) were strength-trained for six days. Control mice failed to show any appreciable increase in muscle fiber size after this short period, while, in contrast, the testosterone-exposed mice showed a 31 percent increase!

This phenomenon was known, but not understood, by people who did weight training. Ingrid Egner and her crew explained it. They said that previously untrained muscle fibers recruit nuclei from activated satellite (stem) cells before growing larger. The nuclei are the command centers driving the muscle cells to produce more protein to get larger and stronger, and it seems that this greater number of nuclei is retained and protected over time. Muscle fibers with this higher number of myonuclei then grow faster when given an exercise stress like strength training. This "memory" of prior strength apparently remains stable for up to 15 years and may actually be permanent.

So here's the interesting part: strength training early in life might be beneficial later in life. A brief period of anabolic steroid use may cause long-lasting performance enhancements that continue many years after use is discontinued. It is almost as if the "use it or lose it" adage has been changed to "depending upon what you used you might not really lose it." This would render useless short term bans with later reinstatement in sports.

Captain America's super soldier treatment enhances his abilities very quickly, but he may possess the ability to be retrained quickly forever. In a similar way, athletes who use steroids may possibly be permanently advantaged.

We move now from moving fast to thinking fast. Captain America is known not just for his muscles, but also for his tactical genius. Let's peer inside his brain and see what might be optimized using the super soldier procedure.

6. THINK!
PUTTING KAPOW AND KNOW-HOW INTO CAP'S CRANIUM

That was the whole point: replicate the original Super Soldier procedure, give them their army, and then see what's next. You think Steve Rogers was a tactical genius before he became a Super Soldier? Steve Rogers was dumb as a sack of hammers . . .

—Bruce Banner (the Hulk's alter ego) in *Ultimate Hulk vs. Iron Man: Ultimate Human*

If the brain were so simple we could easily understand it, we would be so simple we couldn't.

—Emerson M. Pugh

When I was growing up, in the 1970s, I wanted to be an astronaut. One of my favorite TV shows was the *Six Million Dollar Man*, an ABC production based on the novel *Cyborg* by Martin Caidin. The lead character, Steve Austin, was a former astronaut who got injured in a flight crash and had to be "rebuilt" using bionics.

As part of the rebuild, Austin wound up with bionic legs, right arm, and left eye. These bionic enhancements gave Steve extra strength, the ability to run at 60 miles per hour, and zoom vision. Back then, these cybernetic advances seemed cool but impossibly farfetched. They were in the realm of science fiction. Well, these technologies aren't far away now. They are real and present and more are on the way soon. So what about going for the ultimate cybernetic implant—what about building a better brain?

PERFECT SUPER SOLDIER ENHANCEMENTS BECKONING TO BLACK PANTHER

The first step toward building a better brain turns out to be trying to build any kind of brain at all. On the surface this seems like an easy exercise. The brain is just a collection of special cells, right? So in theory, it should be possible to build a brain by following a few simple steps. First, figure out how the brain functions, then run some computer simulations, use the outcomes of the simulations to create fully detailed models, test and retest the models with machine learning algorithms over many, many iterations, and then make a brain based on the successful outcomes. Job done! Except there are some complications that make this idea, to borrow a bit of engineering jargon, a "nontrivial" problem. That's a problem that's really, really difficult. Engineers have a dry sense of humor.

There is a disconnect between anatomy and physiology that makes understanding how the brain works profoundly challenging. You can tell a lot about how most of your body works from what it looks like. Function (physiology) follows form (anatomy). In your cardiovascular system you've got a big muscular pump in the form of your heart that receives blood from certain vessels and pushes blood all around your body through other vessels. If you took a good look at your heart and thought about all the biological piping coming in and out and the way the valves let fluid move, you'd come up with a reasonably accurate understanding of what the heart does and how blood is distributed throughout your body.

But brain function is different. Your brain contains about 100 billion neurons. Those 100 billion processing cells might have, on average, between 1,000 and 10,000 connections from other neurons. That means about 100 trillion connections, which is a pretty big number. Far bigger than the estimated 200 billion galaxies in the known universe. Overall this is a huge number of connections to model.

This doesn't mean the anatomy of the brain is impenetrable. It is certainly complex, but the general features of the connections from those 100 billion neurons form into tracts and bands of connections within the brain that can be (mostly) identified.

The real nontrivial problem comes from the fact that the function—the physiology and behavior—of the brain cannot be directly predicted from anatomy. Enter those 100 trillion connections. The key thing is that the network performance of the brain emerges from the activity of the synaptic connections that are active at any given time. This activity shifts constantly.

Imagine sitting in a boat that is rising and falling on ocean swells. Boats are all around you and you can see them rising and falling so that at any given moment you see different boats. Those boats all represent active connections between neurons that are expressed when you can see them and silenced when you can't. To complete the metaphor, multiply the number of boats by many trillions. This is what makes building a brain such a daunting task. The challenge is not so much in building something with brain-like connections; rather, it's in building something that functions like a real brain.

BRAINS! THEY'RE NOT JUST FOR BREAKFAST ANYMORE

Okay, so building a brain from scratch is pretty difficult. What about simply amplifying the brain you already have?

Since George Romero's 1968 cult hit film *Night of the Living Dead*, we've known that zombies live and die on brains. But we want to go from zombies with no brain activity—yet a curious taste for humans—to improving brain function as much as possible. How can we make those brains the best they can be? It's generally understood that Sudoku puzzles and brain "gym" activities stimulate your brain. They can help you to maintain a degree of mental acuity, but they can't really make you smarter in the sense of being a more creative thinker. That is, they won't *increase* your intelligence.

To actually make you smarter, you would have to dramatically change how your nervous system—your brain and spinal cord—work. But what if there were ways to do that? To change what Mother Nature provided? Would you choose to do so? And if you did, what would it mean for you as a human? Would you choose to be like Charles Xavier—Professor X of the X-Men—or some other super-smart hero? Black Panther, T'Challa, King of Wakanda and on-again-off-again member of the Avengers is recognized in the Marvel Universe as a genius polymath with a photographic memory. But could Steve Rogers acquire a brain like Black Panther or Tony Stark? It all comes down to increasing brain power, getting smarter, or having more intelligence. For the rest of this chapter, let's explore four approaches that could be used to enhance brain function, to turn Steve Rogers into Captain America, tactical genius. The approaches we'll examine are genetic engineering, stem cell transplantation, machine implants, and pharmacology.

IS SUPER-SMART IN YOUR GENES?

Some people are smarter or more intelligent than others. Maybe there's a genetic component that explains these differences and, if so, maybe we can increase our own brain power by changing our genes. But is there a gene for intelligence, and is intelligence really a heritable trait, something that can be passed from one generation to another? Like many behaviors, intelligence is considered to be a complex trait in contrast to simple traits like hair color. Many of the most powerful scientific studies on genetics and intelligence have used comparisons between identical twins. These have shown that intelligence is, indeed, largely influenced by genetics.

All traits arise because of genetic influences. Despite that, you can't rely 100 percent on your parentage for any particular trait. The crucial point here is that heritability of a given trait— especially a complex one like intelligence—depends on many genes, each of which has a very small influence but together combine to have significant effect.

In molecular genetics, the phrase "genome-wide complex trait analysis" sounds complicated, but the original term, "genomic relatedness matrix restricted maximum likelihood," is much worse. What it describes is a relatively new technique that searches for genetic similarities across individuals in order to predict similarities in actual phenotype. This analysis is performed on a large scale, involving thousands of people across hundreds of thousands of SNPs—single nucleotide polymorphisms—that allows for millions of cross-comparisons. As a benchmark, genome-wide complex trait analysis has shown that despite evidence of a clear genetic influence, only 1 percent of phenotypes such as height and weight can be explained by a single gene. The unexplained 99 percent are called "missing heritability," a reference to the gap between inheritance and the explanatory power of variation in DNA.

In Chapter 4, on shape, we talked about the role for genetics and inheritance in shaping our capabilities. But there's an additional wrinkle with intelligence. It turns out that heritability of intelligence—the amount that genetic factors contribute—goes up as we get older. This "genetic amplification" likely happens because of our interaction with the environment. In a rethinking of nature versus nurture, genetic differences are magnified during development and aging as we choose, change, and create environments that match our genetically programmed dispositions.

This genetic drive has a powerful effect on who we choose to have children with. It matters far more than height, weight, or personality, when, for example, we look for a mate. Verbal intelligence is actually the largest factor influencing our choice of partner. All the twin and genome-wide complex trait studies highlight a key feature of intelligence: genetic influences are general across many diverse measures of cognitive ability. This means that genetic influences are generally for abstract concepts like thinking instead of for specific traits like verbal fluency or mathematical ability. These findings help guide the approaches taken by scientists as they target and study genes that have widespread effects.

ARE THERE SPECIFIC GENES FOR INTELLIGENCE?

Ian Deary and colleagues at the University of Edinburgh are interested in the genetic foundations of human intelligence. Many studies have identified "candidate genes" that may help predict superior intellectual capacity, and this is an emerging and rapidly changing research area. Some of those genes and what they do are:

- oxytocin receptor (OXTR)—affects general intelligence

- insulin-like growth factor 2 receptor (IGF2R)—mathematical ability

- apolipoprotein E precursor (APOE)—related to cognitive function and memory; associated with dementias

- brain-derived neurotrophic factor (BDNF)—affects memory

- neuroplastin (NPTN)—general intelligence

We know these genes are associated with various features of the complex trait of intelligence. We don't have much information on the specific features of neural function that might directly underlie these features. There are some studies showing relationships between SNPs and differences in brain anatomy that may lead to important advances in understanding intelligence and intellectual deficit.

The thickness of the cerebral cortex—the outer part of the surface of your brain, where most of your neurons live—is an indirect measure of the complexity of interactions between the neurons that live there. This cortical thickness has a genetic basis and in children and young adults is strongly related to intellectual ability. In 2015, Sylvane Desrivières from King's College, London, along with collaborators from around the world, showed an important relationship between the NPTN gene (related to neuroplasticity), complexity of connections in the brain (measured as cortical thickness), and intellect.

They chose the NPTN gene because of its role in regulating the development of neurons and the cellular connections between them, all of which means increased processing ability and associated intellectual ability. Uncovering a specific gene that links cognition to cortical thickness takes us closer to understanding the genetic basis of intelligence and the ways in which genetic dysfunction may affect intellectual disability.

Until we better understand the interaction and influence of specific genes of single nucleotide polymorphisms, we can't rely solely on molecular biology in our quest to enhance intelligence. So what about directly augmenting the function of the brain by biological implantation and technological interface?

TISSUE ENGINEERING IN THE BRAIN AND STEM CELL CHIMERAS—HYDRA BEWARE!

A chimera is an animal whose biological material has been combined to create a "new" animal. It traces its lineage back to Greek mythology (the original chimera was part lion, snake, and goat) and the Lernaean Hydra. That multiheaded sea monster posed a particular challenge to heroes who would fight it because it regrew appendages as they were cut off. No such creature exists that can emulate the ancient myths, but chimeras do occur naturally. For example, there are chimeric budgies, which are essentially two separate birds fused into one. Also, in humans, a nonidentical twin may have up to 10 percent of cells absorbed from their sibling.

In the *Iliad*, Homer described a chimera as an animal made from bits of other animals. Nowadays, biomedical chimeras are animals constructed with cells from different organisms. In

2013, Xiaoning Han and colleagues created a latter-day chimera in the laboratories of Steven Goldman and Maiken Nedergaard at the University of Rochester Medical Center. They subsequently speculated on the possibility of augmenting the neural processing ability of one species by surgically transplanting cells from the brain of another "more advanced" species. The recipient species was a mouse, the donor cells came from human tissue, and the result was a mouse-human chimera.

Han and friends wanted to see if they could find compatibility in the mouse brain with certain evolutionary adaptations that human brains have acquired. Brains and computer processors are similarly complex. The frequency response capability of the chip—its "clock speed"—determines its overall processing speed and power. When you take a survey of the brains of different animal species, what's really interesting isn't so much that the neurons differ, but rather how fast the supporting cells in the nervous system, known as glia cells, can function—in other words, their clock speed.

The particular kind of glia that have a role in regulating the processing capacity of the neurons are astrocytes. Calcium is a critical modulator of neuron function because it affects excitability and neurotransmitter release. A key function of astrocytes is to support the activity of neurons by regulating calcium signaling. Human astrocytes propagate calcium waves almost three times faster than those in the mouse brain. (I realize this factor seems almost disappointingly small—are we only that much smarter than mice? But the truth is that a 300 percent difference between species is actually huge.) Han and colleagues wondered what would happen if you surgically engrafted human glial progenitor stem cells—cells that would become glia after transplantation—into the forebrain of a mouse. This involved transplanting fetal brain cells into immunosuppressed mice and then measuring their viability, function, and behavior.

Not only did the engrafted human glial cells survive in the mouse forebrain, but they thrived. The cells' ability to propagate calcium signals as found in the human brain were similar. The next bit of the experiment was important—could the glia also affect cellular processes like long-term potentiation (LTP), a critical cellular mechanism that underlies learning and memory formation? Again, the answer was yes. Roughly a 10 percent increase in LTP was measured in the chimeric mice.

Whether all of these adaptations could actually affect behaviors in the mice was the biggest question of all. The researchers assessed the chimeric mice for fear conditioning, maze learning, and their ability to locate new objects in the environment. Amazingly, the mice showed improved performance in all the behavioral tests that were administered. All the evidence suggests that they were made smarter as a result of the implantation of human brain cells.

Later, Martha Windrem and colleagues at the same laboratories used an expanded protocol to examine the long-term effects of engrafting human glial progenitor cells into the forebrains of neonatal mice. Basically they wanted to know if the transplanted cells would continue to thrive over time, or if they would die away. The results were quite dramatic: there was a steady decline in mouse cells and an increase in the human cell content in the mouse brain. This proportional shift was so strong that after one year the glial progenitor cells found in the mouse forebrain populations were almost entirely of human origin. The implanted human cells "outcompeted" and eventually replaced and "infected" the host mouse cells—an utterly unexpected outcome. The transplanted cells actually took over the mouse brain.

Chimeras have huge potential applications in biomedical research. In 2016, Insoo Hyun, professor of bioethics at Case Western Reserve University in Cleveland, wrote in *Nature* that chimeric-embryo research "has a vital role in basic and translational stem cell science." Jun Wu and colleagues, in the same issue, suggested that while there are many "scientific, medical, ethical, political, financial, and other challenges, and not everything that can be done should be done, we owe it to future generations . . . to proceed."

SHOULD CAPTAIN AMERICA HAVE A LITTLE IRON MAN IN HIM?

Let's suppose we want to get your brain to interface with a machine. The first step is to obtain information about brain activity. The least invasive way to do this is to use electroencephalography (EEG). The electric field potentials that your brain generates are measured by electrodes placed on your scalp. Your brain activity changes depending on what you are doing, but with filtering and software analysis, EEG can be used as a control signal for computers and robotic devices, creating a brain-machine interface (BMI).

Neurosurgeons have used BMI devices to help people with neurological diseases, such as the terrifying amyotrophic lateral sclerosis, or ALS (also known as Lou Gehrig's disease). In ALS the lower motor neurons in the spinal cord that activate muscle gradually cease to function and die. This slowly but steadily weakens the individual, who is eventually completely paralyzed, surviving until they also lose control of the diaphragm for breathing. During this process, though, the neurons in the brain are relatively unaffected and continue to function pretty much the way they always have.

Neuroprosthetics can to some extent compensate for the loss of muscular control by connecting the nervous system to functional assistive devices. For example, retinal and cochlear implants can enhance vision and hearing, while spinal cord implants can relieve pain and increase bladder control. Neuroprosthetics that use electrodes in or above the surface of the brain to detect thought are being developed to help those with paraplegia and quadriplegia, and others with severely limited movement, to control robotic limbs.

Individuals who are early into their ALS can learn to use EEG signals to control a computer device that gives them the ability to communicate for as long as possible. Jon Wolpaw and colleagues at the Wadsworth Center of the New York State Department of Health in Albany developed a BMI system based on EEG activity recorded from the scalp and using a simple stretch cap populated with electrodes. This setup is portable and can be situated in the user's home. Some people with ALS who are approaching complete paralysis are using the Wadsworth device.

Can this technology be usefully applied to enhance function in those who don't have neurological damage? Jon Wolpaw pointed out something that is key to integrating a machine and an organ as complex as the human brain. As Wolpaw first related to me in *Inventing Iron Man*, it's about obtaining consistent and reliable control. Methods using electrodes positioned inside the brain or the noninvasive alternative—with electrodes on the scalp—can establish control that is "really good one day—or one three-minute period or even one trial—and really bad the next," Wolpaw says. He suggests we need to think of machine interface as "skill development"—except that it's developed purely through brain activity.

Wolpaw argues that the real issue is that this kind of "skill doesn't become as consistent as muscle-based skills typically do," likely because this is "profoundly abnormal. The central nervous system evolved and is shaped throughout life to control muscles." So far, most BMI researchers have designed tests in which participants perform just one task at a time. Doug Weber of the University of Pittsburgh School of Physical Medicine and Rehabilitation, another leading scientist in this area, says this is "pretty artificial, since we rarely focus so much attention on simple motor tasks," and that real-life "multitasking would require at least a portion of the BMI control to be performed subconsciously."

Doug Weber has extensive experience in this area. He works to create devices that interface with the nervous system to augment or restore functions impaired by injury. The ultimate application as he sees it is to "enable direct neural control and sensation of prosthetic hands, arms, and legs—think Luke Skywalker in *The Empire Strikes Back*." This approach is also used to understand and modify "neural codes that signal physiological status of immune, cardiovascular, metabolic, and other systems. Many diseases affect the function of these autonomic systems and neural interface technologies offer new opportunities for delivering therapies that restore healthy functioning."

Weber believes that huge benefits will follow from "creating truly new treatments for auto-immune diseases and mental health disorders. Both of these are complex issues that affect large numbers of people. Pharmacotherapies are largely ineffective and usually carry nasty side effects. I believe that direct interfaces with the nervous system offer a more powerful and tunable approach to treatment, leveraging our internal physiology to monitor, diagnose, and treat disease—no drugs required."

There are caveats, however. Although he personally considers it unlikely, Weber explained that extrapolation of his work could lead to "measuring and decoding any signal from the nervous system—maybe even without someone's knowledge or consent, which would be a terrible violation of privacy. Perhaps even worse, one could imagine controlling someone's nervous system—with or without consent—altering their health status or even controlling their thoughts or actions." This echoes a point that Dan Ferris made when he told me that a technological interface that is in the body and "networked" allows the possibility of unwanted hacking "into bionic body parts and holding your body for ransom from afar."

People using these interfaces because they have a severe clinical condition (such as ALS) will accept limitations (such as loss of privacy) that users who are looking for enhancement of normal functioning will not accept. So it's a good thing that, while human clinical applications continue to emerge and reveal further limitations that need to be addressed, basic science continues at a breakneck pace. As recently as March 2016, Sara Reardon reported in *Nature* preliminary evidence that transcranial brain stimulation during training could actually improve jumping performance in skiers by reducing fatigue.

BRAVE NEW WORLD OF BRAIN-MACHINE INTERFACE

The interface of physiology and engineering represented by BMI has a history of successfully restoring function in humans and other animals. This progress was also reflected in the comics in the guise of Professor Arnim Zola. Zola played a key role in *Captain America: The Winter Soldier* movie in 2014 and was listed in the 2016 book *Captain America: The Ultimate Guide to the First Avenger* as a pioneer of "robotics and genetics." He creates a true fusion of biology and biotechnology when his body is laid waste to by disease. In the 2014 film, Zola—or at least the computer representation of him—tells Steve Rogers and Natasha Romanov that "science could not save my body . . . My mind, however . . . That was worth saving. On 200,000 feet of data banks. You are standing in my brain."

With each real-life success in brain-machine interface, though, new questions arise. Can we go beyond just measuring brain activity to control a device to imagining a connected device that can assist in learning? Theodore Berger and colleagues showed that a neuroprosthetic interface enhanced memory function of the rat hippocampus and could overcome memory deficits mimicking natural damage such as occurs in dementia. Similarly, rhesus monkeys have shown improved decision-making by using a neuroprosthetic connected with the prefrontal cortex.

These ideas move beyond simply a control system that can spell words on a screen to a system that can function like part of the brain itself. Simeon Bamford and colleagues established a proof of principle for direct applications of brain-machine neuroprosthetics in motor learning using cerebellar motor control and learning circuitry. The system they're exploring would send and receive inputs from the brain to control devices that supplement the function of the brain itself—for example, the brain's ability to learn.

In 2012, Ivan Herreros and colleagues successfully connected an external circuit to the brain of an anesthetized rat as a "step toward the development of neuro-prostheses that could recover lost learning functions in animals and, in the longer term, humans." Most recently, this technique was used to establish an interface with the rodent brain for testing closed-loop motor learning in real time. This potentially lays the groundwork for refining future neuroprosthetics. Again, this approach, while initially designed to supplement function in a damaged brain, could conceivably also be used to enhance function in a normally functioning brain.

In a mind-boggling study (seriously, if this isn't mind-boggling, nothing is), Miguel Pais-Vieira and colleagues extended this brain-machine interface concept to brain-to-brain interfaces for shared information processing. Two rats had electrode arrays implanted into the sensorimotor areas. One rat served as an "encoder" of sensorimotor information during performance of either a tactile or visual task. The brain activity generated in the encoder rat (at Duke University) was monitored and then relayed to a second "receiving" or "decoder" rat, located in a distant laboratory (in Natal, Brazil).

The brain of the decoder rat was electrically stimulated through the implanted electrode array based on the timing and pattern of activity received from the encoder rat. The behavior of the decoder rat was directed by this activity: it made task choices similar to those made by the encoder rat. In effect, the distant decoder rat was taught by the brain signals generated by the encoder rat, signals that were relayed by the direct brain-to-brain coupling in an "artificial communication channel." In a protocol that could have been plucked from a comic book, this result shows that rats linked through brain-to-brain electrode arrays can learn complex, cooperative, goal-directed behaviors.

It gets better. A related human test of brain-to-brain interaction by Charles Grau and colleagues used EEG to detect signals at the "source" brain (essentially the encoder rat above) and transcranial magnetic brain stimulation for transmission to the "receiver" brain (the decoder rat analogue) to establish that direct communication between the brains of conscious humans is possible. This study focused on transmitting simple language, but it heralds the future arrival of more complex communication.

Mikael Lebedev, a neuroscientist at Duke University working extensively in the field of brain-machine interface, believes that the next century will see amazing developments in sensory, motor, and cognitive interfaces and a huge increase in overall "bandwidth" that we can use. Lebedev told me that solving "medical problems is the major anticipated outcome to help repair neural circuitry damaged by injury or disease." The most impressive future development with sensory interfaces will probably be "restoration of vision to the blind, but also there will be more exotic developments like augmentation of sensations and addition of new senses. This is all very important because sensations are the primary factor in any behavior, and many behaviors are sensation-seeking behaviors."

Lebedev also foresees major advances in brain-machine interfaces for movement. He says "systems for voluntary control over movement of devices directly by brain activity will finally achieve good accuracy. . . . Once good accuracy and versatility is attained, we will see an emergence of 'telekinetic' communications with the outside objects: mind-controlled cars and other utilities." In the nascent area of cognitive interfaces, Lebedev predicts that "mind-reading devices will take off. Say, converting your thought into a text will become feasible. And many other remarkable possibilities. As a result of these developments, humans will receive many extensions of themselves, which they will control as current humans control their bodies."

These are amazing discoveries. All of these could be used to enhance Steve Rogers. Ultimately, they are likely to change dramatically the boundaries that constrain our ability to modify human brain function. Lebedev suggested that all these advances are steps along the path to "losing 'human identity.'" Humans will be literally merging with machines, an outcome that some people may consider regrettable." So what about just cranking up the function of the brain you have now?

WHAT ABOUT A BRAIN BOOST
THROUGH PHARMACY?

It's commonly (and erroneously) believed that most humans, most of the time, use just 20 percent of their brain. In the 2011 hit movie *Limitless*, Bradley Cooper's character, Eddie Morra, takes a drug that allows him to use 100 percent of his brain's capacity. Putting aside for now how silly the myth about brain activity actually is, what if you could take a drug that made you smarter?

This seems like an entertaining science fiction premise, but there really is a vast and burgeoning pharmaceutical industry focused on the "neuromodulation" of thought and memory. These involve the use of new drugs called cognitive enhancers or nootropics. Their development hinges on the idea that brain function can actually be enhanced. Which in turn leads one to ask, why isn't our brain function already at peak performance?

Gary Lynch and colleagues at UC Irvine point out that most of our higher-order thinking comes from parts of the cerebral cortex that have seen the largest evolutionary changes in the past two million years. They argue that it seems reasonable to assume that cognition is not fully optimized in humans. This means there's room for both continued evolutionary changes and also deliberate enhancement. But enhancing in what way? Should we aim to enhance the "natural" efficiency of the brain, or should we make actually gaining new abilities our goal?

The drug Ritalin (methylphenidate), widely prescribed for the treatment of hyperactivity, has received a lot of attention both in scientific studies and from the general public. It has a clear effect on arousal and seems likely to affect performance in thought, but it doesn't seem to do all that much for so-called higher-order thinking. Modafinil (or Provigil) similarly seems to affect arousal levels and improve performance. It was first used as a drug to offset the uncontrolled sleeping behavior that occurs in narcolepsy. But the effects of both Ritalin and Modafinil really seem more similar to caffeine and nicotine than to any true enhancer of brain function.

Ampakines get their name from the ion channel receptor they bind to—the AMPA receptor. This receptor is activated by glutamate, the major excitatory neurotransmitter in the nervous system. Ampakines improve learning, memory, attention, and alertness without the side effects associated with caffeine and amphetamines. Ampakines boost the normal action of glutamate, which leads to long-lasting increases in transmission. That is to say, they help make brain function more agile—a form of neuroplasticity. Think of it as a fuel additive that makes an engine run more efficiently and with more power output. But the engine is your brain and you are both car and driver.

Another class of brain stimulants may change brain metabolism and structure. These include a compound called NSI-189. This compound may help with memory and cognition by stimulating neurogenesis—the production of new neurons from stem cells found in the hippocampus, the major memory storage center deep inside your brain. Dysfunction in the hippocampus is related to post-traumatic stress disorder, major depressive disorder, and Alzheimer's disease. Clinical trials will be needed to figure out how useful NSI-189 may be for these disorders.

All of the drugs we've been discussing have been tested in laboratory animals and shown to be safe and effective. But they haven't all translated that well to human applications. There are a number of possible explanations for the disconnect. Differences in connectivity between the human brain and the brains of other animals, for example, and the lack of tests suitable for use in humans. It is clear, however, that many of these compounds can help with degenerative brain function in Parkinson's disease and dementias.

But what would happen if you took a drug for cognitive enhancement when you didn't have any underlying pathology? To answer that question, I interviewed someone who prefers to remain anonymous—let's call this person Agent 13—who used NSI-189 "recreationally" for eight weeks. Agent 13 had no known underlying neurological disease or disorder and in fact works as a leader in a job that requires constant creativity and higher-order thinking.

EPZ: Why did you decide to take NSI-189?

AGENT 13: I just wanted to give it a try. I'd heard about it and looked it up on the internet and had done my background research on what it might do. The animal studies showing large effects on the hippocampus were compelling to me.

EPZ: What dose did you use per day and for how long? Where did you get it from?

AGENT 13: I got it through a guy I met who worked at another company. He got some from a supplier off the internet and had it independently tested in a lab to make sure it was what it was chemically supposed to be. We read about a range of dosages used from 10 to 150 mg but that the most effective dose was maybe around 40 mg. So we used that dose. The guy I got mine from tried a whole range and didn't find any additional benefits above about 40 mg.

EPZ: How long did it take to have an effect, and what did you notice first?

AGENT 13: It has an effect after about 20 minutes and the biggest thing I noticed first was an overall alertness, a kind of sharpening of the senses. This happened every time I used NSI-189, which was twice a week (Friday and Saturday nights) for eight weeks.

EPZ: What were the beneficial effects?

AGENT 13: In addition to the alertness, I also had a more comfortable feeling around people, and an overwhelming desire to learn things. It also made me feel like I was having sharper memories and maybe remembering things I had forgotten before.

EPZ: Did the NSI-189 stimulate desire for any particular kind of learning?

AGENT 13: Well, basically just to take in knowledge. But I also had a powerful urge to learn how to play guitar. So I went out and got a guitar and learned. I'd never played guitar before but I just had this strong urge to do it and just went and figured it out.

EPZ: How long did that feeling last after you stopped taking NSI-189?

AGENT 13: Well, it never really did stop. Even three months after my supply ran out, I still had the impulse to keep learning and pushing my limits of information and knowledge.

EPZ: Were there any unwanted side effects?

AGENT 13: After about 12 hours, you wind up with a kind of drowsy feeling where you just feel a bit off and spent. Not quite like a hangover, but kind of what you feel like if you've taken too many prescription sedatives and the effects are lingering. Also, although you eventually feel more comfortable in groups, during the initial three–four hours after taking NSI-189, I kind of felt more inside my own head. Like a feeling of isolation even when around others. Sort of a feeling like there was something more going on inside my own head that was isolating me a bit.

EPZ: Would you use this or a related substance again?

AGENT 13: For sure I would. It definitely made me feel sharper. My creativity was really enhanced, and I was thinking of going further in my own work and feeling like I could push more outside my earlier comfort zone.

EPZ: Do you think of this as a form of human enhancement—maybe a kind of brain doping?

AGENT 13: Yes, I think it would be fair to think of it like that. My brain was definitely enhanced by the NSI-189. It certainly increased my creative ability and the sharpness of my thinking. But I can't see the down side that usually comes with a word like "doping." I wasn't in any kind of competition with rules or breaking any laws by using it, so if it's brain doping, it's not an ethical issue. I am trying to find a supplier for more because I like the effect it has on me. But if it were an illegal substance, I wouldn't use it. Another thing about the concept of doping. When I was in my sporting career I had lots of coaches tell me that I would have to take steroids or other stimulants if I wanted to take my game to the next level. And there are lots of side effects with steroids. Here there weren't any really major side effects. This kind of "doping" and probably all other types, if they aren't used to break rules in sport or laws in society, should be a personal choice for each one of us to take our abilities as far as science can help us.

I asked Dr. Peter Reiner, professor and co-founder of the National Core for Neuroethics and a member of the Department of Psychiatry and the Brain Research Centre at the University of British Columbia, to comment on Agent 13's experiment. Reiner's work on neuroethics focuses on public attitudes toward cognitive enhancement. Peter told me

> there are several things interesting about this exchange. The first is the notion that people are finding out about experimental drugs online and then accessing them for cognitive enhancement purposes before they are proven to be safe and effective for treating disease. This is similar to the movement in the world of "do it yourself" brain stimulation where early adopters are willing to take on risks—some would say undue risk—in order to enhance themselves. More than anything, it seems to me that this is a comment on the value of cognitive function in modern society.
>
> For all of these individuals are trying to improve their ability to compete in the modern world, as you very clearly articulate when making the distinction between cognitive enhancement and steroids in sport.
>
> The question is, where is this going? It seems to me that the prospect of pharmaceuticals being developed that are truly

effective is a long shot; there have been efforts underway for over a decade, but the process of drug development—especially for modifying the brain—is long and tortuous.

There is a higher likelihood that one of the various ways with which one can stimulate the brain without breaching the skull—transcranial direct current stimulation and its kin—will achieve success sooner, especially as the development cycle is shorter. And finally, it seems as if the prospect of using technological devices to enhance our cognitive abilities is even more likely to emerge as an important means of cognitively enhancing oneself in the near future.

BETTER BRAINS WITH ARTIFICIAL INTELLIGENCE

Thinking (sorry) about brain enhancement with technology brings to mind (again, sorry) that we have to also consider the likelihood that technology itself can be enhanced by artificial intelligence. I'm thinking of technology that is all around us and growing in extent. You soon might have a robot vacuum cleaner chugging around your kitchen or a robot lawn mower auto-trimming your lawn. The continual evolution of AI and its increased presence in our world speak to achievements in science and engineering that have tremendous potential to improve our lives.

Or possibly destroy us and everything we hold dear. This theme figured prominently in the Marvel Studios *Avengers: Age of Ultron* movie in 2015. In that movie, Ultron represented AI at its worst and consequently ran up against Captain America and the Avengers at their best. The theme is a timely one.

But let's pause to consider how bad AI could be, using Ultron as an example. Well, according to the *Official Handbook of the Marvel Universe*, Ultron is a "would-be conqueror, enslaver of men" with genius intelligence; superhuman speed, stamina, reflexes, and strength; subsonic flight speed; and demigod-like durability. Interestingly, and perhaps surprisingly, the handbook also says—clearly on the good news front—that Ultron has only "normal" agility and "average hand to hand skills." I think this means that if you can get up close to a huge robot with artificial intelligence and superhuman speed, you should have a (brief) fighting chance. Okay, maybe not you or me, but Captain America!

In Marvel comics and movies, Ultron seeks to overthrow the humans who created it. On the one hand, Ultron seeks to bring peace and order to the universe. (Cool!) On the other hand, this peace and order can only be achieved by eliminating all other intelligent life in the universe. (Not cool.)

Age of Ultron is a great foil for exploring the fictional conflict between biological beings and artificial intelligence. But how fictional is it really right now? We can find answers in the fields of machine learning, artificial intelligence, and artificial life.

In 2015, Google DeepMind's Volodymyr Mnih and his collaborators challenged a neural network to learn how to play video games. They wanted to find out how good the software could get at learning from one game and transferring that mastery to another game. Their artificial neural network was designed to have roughly the same skill as a "human" gamer. By the way, the games chosen were true 1980s classics, such as *Boxing*, *Video Pinball*, *Robotank* (a favorite of mine), and *Tutankham* (another favorite).

Although learning to play a video game might seem a trivial accomplishment, in the bigger picture, Mnih and co. were doing serious work. They were setting out to demonstrate that an AI system could learn and adapt its skills to situations for which its original programmer had never prepared it. This application of new skills in new ways is the hallmark of independent thinking. Which is cool, but also scary.

WHAT? ME WORRY?

Media coverage of AI is rarely positive. There is a boundary that separates undeniably helpful applications of AI—imagine a medical application where robot-conducted surgery is performed in a remote community and overseen by a physician in a distant location—from truly frightening scenarios of near-future military conflict. It doesn't take much imagination to see how the combination of current combat drone technology and AI computer engines could lead to independently directed machine warfare.

We don't always recognize these kinds of boundaries until we find ourselves on the other side, in a strange land, fully committed to something we don't know how to control. In science we often push to discover and apply things before we truly understand all the implications—both positive and negative—that flow from our creations. In 2014, Tesla CEO Elon Musk told an MIT symposium, "I think we should be very careful about artificial intelligence. If I had to guess at what our biggest existential threat is, it's probably that." In January 2015, Musk then put real money behind his concerns when he donated $10 million to a nonprofit focused on AI safety.

Pedro Domingos explored this theme in his book *The Master Algorithm: How the Quest for the Ultimate Learning Machine Will Remake Our World*, writing, "people worry that computers will get too smart and take over the world. The real problem is that they're too stupid and they've already taken over the world." No less a geek than Bill Gates revealed his reservations about AI in a Reddit "Ask Me Anything" session. Gates wrote that "I agree with

Elon Musk and some others on this and don't understand why some people are not concerned." Stephen Hawking coauthored an article on the risks of AI, saying it could be the "worst mistake in history." So, before we go too far down that road, maybe we should make sure we have the Avengers—led by Captain America, naturally—at our side, ready to spring into action.

MAYBE WE SHOULD TRY LIVING TOGETHER FIRST

Clearly, knowledgeable people worry about the implications of artificial intelligence. They worry especially about its potential independence from human interaction and oversight. To avoid such an outcome—and that sci-fi end-game of Ultron—maybe we should adopt the "collaborative intelligence" approach advocated in 2015 by computer scientist Susan Epstein. Epstein wrote that we traditionally build machines because we need help. We often go for all the bells and whistles, but perhaps a less-capable machine could be equally helpful. By allowing humans to do things that we're better at anyway, such as pattern recognition and problem solving, we could work with machines to do better jobs overall. By deliberately building limitations into our intelligent robots, we could allow them to perform their jobs while simultaneously keeping them in check.

Daniel H. Wilson, in an essay entitled "Dude, Where's My Jetpack?" in *Discover* magazine, echoed the early sci-fi writers Jules Verne, H.G. Wells, and Isaac Asimov when he described a future that is "supposed to be a fully automated, atomic-powered, germ-free utopia." This way of thinking about AI implies that humans can employ and work with machines, rather than being replaced by them. This collaborative view, however, is at odds with the traditional industrial imperative that calls for the mechanization of operations wherever and whenever we can. We can only hope, for the sake of our future selves, that these real-world conversations will continue as we progress toward smarter and smarter machines.

ISAAC ASIMOV TO THE RESCUE (AGAIN)

Isaac Asimov, one of my favorite science fiction writers, anticipated our present concerns about AI back in 1942. In his influential short story, "Runaround" (later collected in the book *I, Robot*), he presciently observed that when a human operator can access the functional capacity of a machine, the reverse is also true: the machine likewise has access to the functional capacity of the human. If future neural interfaces function indistinguishably from the user (since they are part of the control system that manifests as the will of the user), it may be impossible to separate the actions of the neural interface from that of its wearer. Such an outcome places a higher order of responsibility on those "augmented" users. Accordingly, Asimov laid down his Three Laws of Robotics, which aimed to protect the sanctity of human life.

I've reworked Asimov's laws to apply to the emerging complexities of machine-brain-machine interfaces. An augmented user with a neural interface

- may not injure a human being or, through inaction, allow a human being to come to harm (Law 1)

- must protect its own existence as long as such protection does not conflict with Law 1 (Law 2)

Note that my revision has the effect of deleting Asimov's original Law 2. That law stated that a robot "must obey orders given it by human beings." It's irrelevant because, in the future we now envision, the user is the interface and the term *human being* applies to all and related subspecies. In 2016, Jim Davies, professor at the Institute of Cognitive Science at Carleton University wrote in *Nature* that, "rather than obsess about consciousness in AI, we should put more effort into programming goals, values, and ethical codes. . . . the first superintelligent AI will be the only one we ever make."

It's to future humans—and how long their future can last—that we turn next. After all, a major outcome of the super soldier treatment was to delay the aging process in Captain America.

7. LONGEVITY!
THE STEVE ROGERS REGENERATION AND RETIREMENT PROJECT

Dr. Erskine said that the serum wouldn't just affect my
muscles. It would affect my cells. Create a protective system of
regeneration and healing . . . which means, I can't get drunk.
—Steve Rogers speaking to Agent Peggy Carter,
in *Captain America: The First Avenger*

I still have the energy I had at fifty . . . More.
Where is it coming from? Honestly, I don't know.
I wish I knew. It's a mystery even to me.
—Track star Olga Kotelko at age 91, in *What Makes Olga Run?*

It wasn't part of his original suite of "powers," but Captain
America eventually gets a healing factor from his super soldier
treatment. Imagine being unbreakable with the ability to do
anything you liked and not get hurt! The gift of invulnerability
has been a great attraction for many fiction writers. Superman's
ability to withstand (almost) anything, along with being able to

fly and shoot rays from his eyes, is a key part of his mythology and appeal.

While there isn't anything on the real science front that can make you invulnerable, there are some things that can help with perhaps the next best thing—healing fast. The ability to rebound from injury has played a part in just about every vampire or werewolf book, movie, or graphic novel. And there's a link between healing and aging in our cells. Quick healing ability is also a main part of the attraction to the Marvel Comics characters Wolverine and Deadpool, as well as Captain America.

WOLVERINE AND SHEDDING LIGHT ON HEALING FASTER

Wolverine, developed by the team of Len Wein, John Romita Sr., and Herb Trimpe, debuted in *The Incredible Hulk* #181 in November 1974. Like many comic book heroes, he has multiple origin stories. In the 2009 graphic novel *Wolverine: Origin*, we meet Wolverine as James Howlett living in mid-1880s Alberta, Canada. He then moves to a mining town in northern British Columbia and goes by "Logan." After leaving town and spending some time living with wolves and foraging from the land, Logan comes back to live with the Blackfoot Indians before entering the Canadian army in World War I. Later he moves to Japan and, later still, in World War II, hooks up with Captain America as a kind of mercenary for hire. He serves in an elite Canadian unit, and then makes his way to the CIA and finally becomes part of "Team X." Black ops.

Wolverine's memory is forcibly altered by Team X scientists, but he eventually breaks free of their control to rejoin the Canadian Defense Forces. Then he gets kidnapped by "Weapon X." It is here in the story that the adamantium unbreakable alloy is grafted to his bones. The alloy complements his biological mutant abilities so that his main power is rapid healing or "mutant healing factor." This capacity for regenerating damaged, or even destroyed, bodily tissues exceeds by far anything a "nonmutant" is capable of.

In his various depictions, Wolverine seems pretty much indestructible, almost like Superman. The difference is that Wolverine does get hurt and injured. A lot. But he heals so fast it hardly seems to matter. This is great for Wolverine. Is comparably rapid self-healing a possibility for humans?

Healing and cellular repair after injury are apparently linked to the ability to regenerate whole parts of the body and this, in turn, is related to the process of aging. Animals like the axolotl—or *Ambystoma mexicanum*, a species of salamander—has attracted the interest of scientists because it can regenerate whole limbs. It also accepts tissue transplants from other animals very readily. In effect, it possesses a naturally occurring version of the mutant healing factor found in Wolverine. Except without the sharp claws.

Because the axolotl and some other animals—including newts and planarian worms—can regenerate their own tissues, some scientists speculated that this remarkable capacity might once have been the property of all animals, a trait buried deep within their DNA. For a long time, it was thought that if we could learn how the newt genome was coding for healing and regeneration, we could figure out how to switch it on in humans.

In one of those unexpected discoveries that make science so exciting, in 2013 Thomas Braun and others at the Max Planck Institute for Heart and Lung Research in Germany showed that the newt's capacity for self-healing, far from having ancient origins, could be a unique and recently evolved specialization. They found, in addition, that the genome itself is much larger than originally estimated, clocking in at an order of magnitude larger than the human genome.

Yet other advances provide more clues. In 2016, Eugeniu Nacu and colleagues showed that two key signaling molecules regulate cellular regeneration—including whole limbs—in the salamander. These molecules (Shh and FGF8) are evolutionarily preserved in many species, and it's thought that applications to mammalian species may be possible.

Well, we aren't quite in a position to regenerate whole limbs yet, and the information we need to get there may not be found in the obvious places. Nevertheless, we have a much better understanding of, and have made progress in, improving healing, by using low-level laser light. For example, Biolux Research's OsseoPulse Bone Regeneration System can speed up healing after dental procedures by as much as 50 percent.

This technology—the power behind the procedure—is actually pretty simple. It is based on photobiomodulation, which means it uses light to change biological function. No gamma rays, X-rays, or any other deadly sounding comic book energy rays need apply. Photobiomodulation uses the energy in the photons that form parts of the visible spectrum of light to activate cellular processes that aid healing. Low-level laser light penetrates deeply into cells and influences the cellular powerhouses—the mitochondria—within the injured tissue. Photons interact with the enzyme cytochrome C oxidase, and this energy is then converted to chemical energy that our cells can use—adenosine triphosphate (ATP). The functional result is increased tissue blood flow and energy supply—both of which are critical for effective healing.

This absorption of light energy and subsequent conversion into chemical energy for use in cells is broadly comparable to the process of photosynthesis in plants. Many scientists believe that the evolution of both plants and animals (including humans) shared paths and energy (from the sun). They speculate that this ability of human cells to absorb light and use it for healing and other functions is a dormant capability, which was replaced over time by a more efficient process of circulation of chemical energy in the blood (from food) and cellular conversion to chemical energy in the cell (ATP). Dormant but not gone, just waiting for better applications of photobiomodulation technology in medicine.

This approach has been used in many different tissues, ranging literally from head to toe and from common applications advertised on late-night TV for recovering from baldness, to exotic applications like acute recovery from heart attack and stroke. In the last few decades, there has been much research into photobiomodulation technology and its various clinical applications. And now there are many medical devices that have received regulatory approval from the Food and Drug Administration and are on the market.

Kevin Strange, president and CEO of Biolux Research, told me that his company has been in the vanguard of this research for the past 15 years. Strange called their new product, the OrthoPulse, a "great example of translational basic research from bench to bedside and beyond culminating in a useful product for patients." Engineers and scientists in many areas of biomedical research look for this kind of transition from research concept to useful real-world applications. The space in which they work is always changing as technology advances and the focus of medical practice shifts. The challenge is to have the technology, medical challenge, and market opportunity converge.

Even though we are nowhere close to achieving the incredible recovery and healing that Captain America is capable of, two things are clear. We are developing new technologies and products that support more rapid healing and more rapid recovery from many forms of injury and disease—outcomes that can benefit everyone from Steve Rogers to you and me. Also, that Vita-Ray treatment Stan Lee described for Captain America's transformation is probably photobiomodulation at its best!

SO MUCH FOR THE MANY HEADS OF HYDRA
IS IT CELLULAR DEATH BY DIVISION?

Biological aging, or senescence, basically starts as soon as life begins. I aged while I wrote this sentence, and you're aging even as you read it. We tend to think of the effects of aging in terms of declining abilities or functions. We begin to show these effects after about the third decade of life. Unless we are Captain America, that is.

Aging is not a matter of a single process but of multiple physical processes. They all have to do with the operation of the cells in your body. Throughout our lives, our bodies maintain a pretty even balance between the two processes of death and division. The number of cells in your body stays pretty constant because you never stop producing more cells through cell division (mitosis). If you had uncontrolled death or growth of cells, you would have pathology. Cancerous tumors are an example of too much growth with not enough cell death. The result can be fatal to the organism as a whole.

How long does a cell live? We know that after some period of time, cells lose the ability to divide, and they die. It's been thought that perfectly normal processes within each cell gradually lead to accumulated damage that stops division. For example, strands of DNA are damaged and the cell fails to flourish or dies via self-destruction. This "apoptosis" is a form of programmed, "deliberate" cell death. Sometimes it's referred to as "cellular suicide." That phrase paints an unnecessarily negative picture, however, because apoptosis is a useful and necessary process. Programmed cell death works as a kind of cellular "clean-up" that benefits the remaining cells in the organism.

Coming back to DNA strands: the idea is that the telomeres need to stay nicely put together. If they fray, contact with other DNA strands can occur, leading to deterioration within the cell. The telomeres work kind of like the plastic bits at the end of your shoelaces—they keep the material from fraying. When that plastic comes off your shoelace (as it always does eventually, let's be honest), the ends of your laces unravel. Those ends cannot be easily tied or even put through the eyelets anymore. In a conceptually similar way, frayed DNA (or, really, shortened telomere length) is an outcome of certain diseases and also occurs in aging. Maintaining an optimum telomere length is therefore important.

It's so important, in fact, that the discovery of the enzyme that maintains the telomere length—cleverly called telomerase—earned the people responsible the Nobel Prize in 2009. Elizabeth H. Blackburn, Carol W. Greider, and Jack W. Szostak won the Nobel Prize in Physiology or Medicine for their "discovery of how chromosomes are protected by telomeres and the enzyme telomerase."

IS THE SOLUTION TO AGING NO SEX, NO DRUGS, NO . . . FOOD?

An organism's life-span is a function of all the processes happening in all of its cells and is strongly related to metabolic activity. All animals have different rates of metabolic activity and have a different maximum life-span.

If we live to be 85 or 90, we are thought of as having a full life-span. The same "elderly" age for a mouse is about three years. Physiological processes in different species are, well, different. For example, there are variations among species in the effectiveness and efficiency with which DNA is repaired. This matters because we know that telomere length gets shorter with age, and anything that can help maintain length would help offset that aging effect.

What if you could live forever? To think about that let's consider the planarian worm, a kind of flatworm. This worm is able to regenerate itself fully even when split up sideways or lengthwise. Importantly, Aziz Aboobaker and colleagues at the University of Nottingham showed that the planarian worm is able to maintain its telomere length during this regeneration. Worms have it going on. Planarians have the ability to reproduce sexually or asexually. Aboobaker and colleagues saw that the worms reproducing asexually had a very large increase in telomerase during regeneration. This means that the health and length of the telomeres are very carefully maintained. In principle, this could produce an immortal cell line for these worms.

Another wild card in the game of long life is the "Methuselah gene." Or, really, the Methuselah genome, a specific combination of genes that allows some people to live into their 90s and beyond. This is also the set of genes that allows some people to smoke packs of cigarettes a day and never get lung cancer. They have some protective interactions between their genes, hormones, and the environment. Creating that matrix for deliberate use is a tantalizing prospect.

In 2012, Paola Sebastiani and others in Boston published a study in which they investigated the genetic factors that contribute to "exceptional longevity in humans." Longevity is another of those "complex phenotypes"—like intelligence, which we discussed earlier—and reflects a mixture of effects, including how we spend our time, what we eat, where we live, and our genetic makeup.

To try and get a handle on the genetic part, Sebastiani and her colleagues did a genome-wide association study of more than 800 people who had lived to be at least 100 years old. They used a complex model of genetics, with an appropriate population of healthy controls, in experiments that found patterns to genetic signatures that were linked to different life-spans.

At this stage of their work, they're unable to say exactly what factors are critical. But they have shown that their method is sound and that exceptional longevity seems to be related to the genetic effects of variants that counter genetic expression of diseases that reduce life-span (disease-risk alleles). Not exactly the Methuselah gene, but it will help future studies with much larger samples of people to narrow the net.

One way to slow down aging, and which fits with the reshaped origin story of Steve Rogers for Captain America in the 1960s, is to slow down metabolism. By going on ice, literally.

FROM CAP ON ICE TO LEADER OF THE PACK
WINTER SOLDIER INTERLUDE

"Put him back on ice," said evil Hydra scientist Arnim Zola. The "him" Zola was referring to was Bucky Barnes in the 2014 Marvel movie *Captain America: The Winter Soldier*.

In addition to their World War II capers, Captain America and Bucky Barnes have something else in common: they have both been frozen. Cap was frozen after the plane crash that ended his first movie in 2011, and then thawed out and returned to active duty in *The Avengers* (2012). Bucky Barnes (aka the Winter Soldier in Ed Brubaker's classic comic book run and

FIGURE 6: Frozen superheroes must have a few things in common with the wood frog. (Image courtesy of Kris Pearn)

in the movie by the same name) has been frozen and thawed many times over the years. This freezing and then thawing thing is a problem for human cells. In order for Steve Rogers and Bucky Barnes to survive the deep freeze, they'd need to share something in common with the wood frog, *Rana sylvatica*.

The wood frog's habitat spans North America from southern Ohio to the Alaskan interior, encompassing a huge range of climates. Samples taken in Fairbanks, Alaska, showed that it can survive complete freezing of more than 60 percent of its body water and temperatures of −18 degrees Celsius. The basic mechanism of protecting their cells from damage is related to the properties of body fluids, particularly a reduction in the absolute freezing point of total body water. This cryoprotection comes about when the concentrations of urea (the main nitrogen-sink found in your urine) and glucose (the simple carbohydrate found in your blood) are increased. This leads to less ice formation during exposure to sub-zero temperatures. It also has the effect of shunting water and thus ice formation to the lymphatic system. This is pretty critical because it reduces damage throughout the frog's body by maintaining the integrity of cellular membranes.

Jon Costanzo and colleagues at Miami University in Oxford, Ohio, wanted to know how the wood frog could survive such low temperatures, even being completely frozen for two months at −4 degrees Celsius, and then be thawed again at 4 degrees Celsius and regain function. Costanzo and colleagues found that in comparison to their Ohioan cousins, Alaskan wood frogs have a much larger capacity for producing glucose. They can survive with much lower glucose levels when thawed and concentrate the cryoprotectants urea and glucose at much higher levels in the brain. There are differences between the Alaskan and Ohioan wood frogs in their ability to survive extreme cold. This means that unique evolutionary adaptive mechanisms exist within the same species, depending on where the frogs live. The same group of researchers later showed that the amount of cholesterol found in the "phospholipid" cell membrane is also much lower in Alaskan frogs.

Which brings us back to *Captain America: The Winter Soldier*. We now know that in order to survive prolonged periods of freezing, it would have been necessary for something more than the Super Soldier Serum. An additional biochemical mix would have been needed to trigger some kind of adaptive process involving glucose levels and urea handling—which would have been rather tricky. We also know now that the process wouldn't necessarily have been restricted to Steve Rogers. Bucky Barnes could also have had the same adaptation. If more than one wood frog can do it, so can more than one World War II comic book patriot.

In addition to stopping cells from aging by freezing them, another Avenger has the ability to alter the aging process. That Avenger is Dr. Hank Pym—Ant-Man!

AN AGING ANT-MAN AND THE NEW SCIENCE OF SENOLYTICS

He was one of the original Avengers. His main claim to fame is his ability to keep his normal human strength when shrunk to the size of an ant. Which sounds like an awesome power, as long as fighting other ant-sized villains is on the menu. Luckily, that was the case in the Marvel Studios 2015 movie *Ant-Man*. Ants might be small, but they are mighty. They can attack and defend not only by biting, but also by stinging with injected or sprayed chemicals. Here, though, the focus is on what happens to Ant-Man when he's ant-sized and how time passes for him when he's small versus big. And I don't mean does time pass in slow or fast motion, but rather, how fast does he age?

The mantle of Ant-Man has been worn by a number of people since biophysicist Dr. Hank Pym discovered his "Pym Particles," which enabled him to shrink to insect size. He also invented a helmet that could control insects so he could have a posse with him and under his control while ant-sized. Of course, Pym also invented Ultron, who subsequently sought the complete destruction of our species.

Professional thief Scott Lang stole Pym's Ant-Man suit in order to save his daughter, Cassie, who had a heart defect. This leads to Pym's reformation—and plays out in the backstory in *Ant-Man* the movie—and Hank Pym becomes a mentor for Lang.

Thinking about how much Ant-Man ages when he shrinks down to his tiny size helps us to get at how we age in our normal everyday lives. Ant-Man was introduced in the comic book *Tales to Astonish* back in 1963, another in the multitude of Marvel characters created by Stan Lee, his brother Larry Lieber, and Jack Kirby.

Dr. Leonard Hayflick started investigating the factors that affected the life-span of cells back in the 1960s. As mentioned earlier, he found that animal cells have a limited capacity to reproduce. He determined that, instead of replicating indefinitely, the cells had a limit—which came to be known as the Hayflick limit—of approximately 50 times.

Earlier we talked about how the life-span of any organism is the sum total of all the processes occurring in all cells in the organism. We noted that animal species each have a different rate of aging and different maximum life-span. Yet the general pattern of aging is similar across species. Humans have a maximum life-span of about 85 years; ants live, at most, for about three years. If we were to scale cellular clocks while watching *Ant-Man*, we would have to keep in mind that for every minute that Scott Lang spends shrunk to ant size he is actually aging roughly 28 minutes in human time. To avoid premature aging at human size, Scott would need something to counteract this effect. There is one possibility: he might be able to alter the mechanisms that regulate metabolism and cellular growth. A key regulator is rapamycin—also known as mTOR—terms that sound like they were comic book chemicals created by Stan Lee himself. Without a doubt, rapamycin would have to be part of the Super Soldier Serum for Captain America.

Rapamycin is named for the antifungal found in soil samples taken from the island of Rapa Nui. It has immunosuppressive and antitumor effects, along with the ability to extend life-spans in yeast, worms, fruit flies, and mice. It's also approved for use as a treatment for numerous cancers and could help reduce the burden of morbidity that befalls most humans as we age. Extension of life-span may lead to a "compression of morbidity," where most of the illnesses across our lives can be compressed into a shorter time frame occurring near the end of life.

The discovery and development of senolytic drugs that could inhibit mTOR—pharmaceuticals that interfere with aging-related breakdown in cellular reproduction—is significant: they could be used to change the human life-span. They may also have applications for altering disease processes such as malignant tumor growth in cancer. In any case, senolytic medications would clearly need to be included in the chemical cocktail Hank Pym gives to Scott Lang for use as Ant-Man.

While science is unlikely to discover Pym particles anytime soon, understanding longevity in biological organisms will provide insight into the mechanism of aging, how to increase life-span, and how to reduce the burden of age-related loss of function.

There's another facet of ant life that gives us some insight into the extension of life-span. *Eusociality* is the term used to describe the highest level of organization of animal society. It encompasses such developments as cooperative care of the "brood," the presence of overlapping generations in adult colonies, and a distinct division of labor. We see this in bees and wasps as well as ants. Importantly, as Laurent Keller explained to me, the "evolution of eusociality has been accompanied by a hundredfold increase in life-span. It would thus be like a primate living for four thousand years." Cooperative living extends life!

It's fitting, then, that this colony-style "team-work" approach found in the real life of ants is also found in Ant-Man's main team-up with Captain America—the Avengers. Of course, any enhancement in longevity that comes with participating as part of that team is probably offset by the dangerous nature of day-to-day life in the Avengers Mansion.

John Morley, professor of geriatric medicine at St. Louis University and coauthor of *The Science of Staying Young*, believes that in the next century hormone supplements and manipulation—particularly anabolic steroids or derivatives—will dominate progress in the biology of aging. Muscle loss—sarcopenia—is a major outcome of aging. Hormones, which increase and maintain muscle strength, have the potential to offset the effect. Breakthroughs in stem cell research may also prove crucial. "Stem cells may turn out to be better at rejuvenating muscle tissue, actually, and those stem cells will most probably come from the person's own fat cells," Morley told me.

"Not only will we be stronger and fitter," Morley said, "we will also be sharper mentally due to advances that reduce or eliminate degenerative brain diseases and dementias like Alzheimer's. . . . Again stem cells may be more efficient." Cell metabolism will be altered by changing the properties of the cellular power stations— the mitochondria. These alterations will also reduce the "oxidative damage" that degrades brain and muscle function as we age. Such changes can be achieved either genetically or in combination with drug therapy. Human physicians may play a secondary role in applying these therapies, as computers with AI capabilities will provide diagnosis and treatment plans in their place. Similarly, computers will be able to diagnose cancer and other conditions using electromagnetic scans.

Exoskeletons and robotic assistants will be the norm, says Morley, and computer microchips in the hypothalamus will boost memory in those with dementia and Down syndrome. He sees the use of these same approaches to enhance "normal" function as a possible problem, however. New therapies that maintain muscle strength in the aged will not necessarily be matched by the ability to control ourselves.

What will happen if a life-span of 120 years becomes the norm? There are precedents. For example, the late Frenchwoman Jeanne Louise Calment, born February 21, 1875, died on August 4, 1997. Such an extended future for the rest of us is conceivable. George Church is a visionary maverick and medical geneticist at Harvard University. When I interviewed him, Church provided some interesting background on this topic. Using the "war on cancer" as a reference point, Church explained that most of the advances since 1971 (when U.S. President Nixon signed the National Cancer Act) have been in the area of prevention. There has been a reduction in smoking. Some known carcinogens have been identified and banned. Presymptomatic surgery based on genetic screening has become more common. The HPV vaccine is in relatively widespread use. There are many other examples. While it's possible that we could just carry on with strategies such as these, alternative strategies could achieve more dramatic results.

Church remarked that diseases like smallpox and polio— once a major scourge—are effectively extinct. At the same time, many biotechnologies continue to advance at an exponential rate that's even faster than that for electronics. All this forms the basis for thinking that we will eliminate many of the diseases carried by vectors like insects and nematode parasites by using gene-drives.

Like John Morley, Church suggests we may be able to reverse aging by reprogramming adult cells, and that "the biggest contribution to our species would be to increase geographic and genetic diversity and adapting to space and other planets . . . where we would need increased radiation resistance, extremely small body size, control of pain (via gene editing), germ-free or pathogen environment, selection for rare adaptations to unusual motion/gravity, circadian disruption, confined spaces." Such far-ranging adaptations could see us "manufacturing essential nutrients in our bodies and a shift in choices to support (with parental training and genetics) a much wider range of neural diversity—for example, by having intentionally adjustable levels of high functioning OCD, ADD, autism, dyslexia, super-recognizers . . . etc."

In the near future, if we "molecular computers" get "good enough at self-modification, we may stay well ahead of electronics (which is currently 10,000-fold less energy efficient) . . . and use very fast and compact molecular storage—in DNA." The last is a nod to Church's own work, where he demonstrated the ability to convert written information into genetic code. In 2015, Church wrote software code to translate a 350-page book—*Regenesis: How Synthetic Biology Will Reinvent Nature and Ourselves*, by Church and Ed Regis—into binary code and then into genetic code using only the As, Ts, Cs, and Gs found in DNA nucleotides. The drop of synthetic DNA Church created was no larger than the period at the end of this sentence. And it contained 70 billion copies of the book!

Let's put some of those applications to the test while we consider the real-life creation of Captain America.

8. CREATING CAPTAIN AMERICA
ENGINEERING A SUPER SOLDIER WITH
SEX, DRUGS, AND ROCK 'N' ROLL

Unlike Superman . . . Cap was not born with his powers.
Instead, he received them through science.
—Thomas Forget, *The Creation of Captain America*

I failed to re-invent the Super-Soldier procedure
that worked perfectly sixty years ago.
—Bruce Banner to Tony Stark, in *Ultimate*
Hulk vs. Iron Man: Ultimate Human

Savvy, strong, and smart. Fantastic fighter, overall amazingly conditioned human, fearless leader. Marvel's Captain America is probably the best of all popular culture examples of re-engineering human form and function to produce superhuman ability. Without question, when it comes to bioengineering, Captain America goes to the head of the class, followed closely by the Hulk. Captain America has had a long career, but he has no supernatural powers. Instead, his abilities arise from that famous Super Soldier Serum, courtesy of Uncle Sam.

In 2014's *Captain America: The Winter Soldier* movie, Steve Rogers arrives on the scene just 5'4" tall and weighing in at 95 pounds. After the procedure, Captain Steven Grant Rogers is 6'2" and 240 pounds. Pretty much the same physical size as that other well-known superhero alter ego, Bruce Wayne. Both Batman and Captain America are pitched as superheroes that are human, but taken to the extreme. Captain America is basically Batman on steroids—and I mean that literally. Plus genetic engineering. And some technological enhancement. Where Batman represents how buff we can make biology, Captain America shows far can we bend biology's rules.

How unlikely a candidate for superhero status was Steve Rogers initially? He wasn't just small and weak. In a scene from *Captain America: The First Avenger* where Rogers is rejected (again, but now for the last time) as "4F," we see his chart in the physician's hands. It lists contraindications such as asthma, scarlet fever, rheumatic fever, sinusitis, chronic colds, high blood pressure, palpitations or pounding heart, fatigability, heart trouble, nervous trouble, contact with tuberculosis, parent or sibling with diabetes, cancer. The physician tells Steve, "Sorry, son . . . you'd be ineligible on your asthma alone." Steve implores the doctor, "Is there anything you can do?"—to which the doctor tells him, while stamping "reject" on Roger's form, "I'm doing it. I'm saving your life."

But what if the power of personalized or precision medicine—where knowledge of a person's genome and proteome are used to guide therapy—could be harnessed to truly make a Captain America?

CHANGING THE GENETIC PROFILE OF STEVE ROGERS

Let's explore what we would need to do to create Captain America using sex, drugs, and rock 'n' roll as our theme. In this revised origin story, to produce Cap we'd need to enhance Steve Rogers—4F and inadmissible to the army—by using

- Sex. It used to be the case that when we were born we got our genetics and the only way to combine and pass on our abilities and characteristics was through sex. Well, that was fun and all, but that's not the only game in town anymore. Say hello to gene doping.

- Drugs. We used to rely only on what we had inside us to regulate our bodies. Now we have access to steroids and other natural and synthetic drugs that can be used to enhance our abilities beyond anything "found in nature."

- Rock 'n' roll. We now have the technology to deliberately change and amplify human capacity through surgery and implantation. New possibilities are emerging in the same way that rock 'n' roll music emerged with the arrival in 1948 of amplified electric guitars courtesy of Fender.

The interplay between all of these is illustrated in Figure 7.

In 2011, genetics pioneer Craig Venter said, "DNA is the software of the cell. . . . When we change the software we change the species." Venter is about changing the germ line. To create Captain America, we'd want to include all the performance characteristics listed in Table 1 to affect muscle strength, endurance, metabolism,

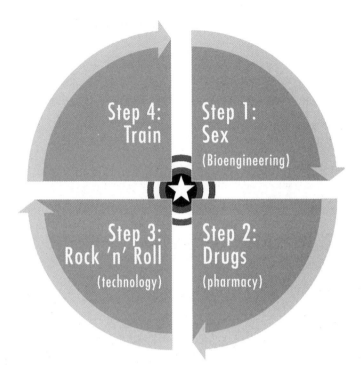

FIGURE 7: Steps in the super soldier procedure.

intelligence, fearlessness, and performance. Gene therapy (or "doping") is the main way we would do this. Genetic engineering really represents the manipulation of the genome through biotechnology or the techniques of molecular biology.

I spoke with Claude Bouchard, professor and chair in genetics and nutrition at Louisiana State University's Pennington Biomedical Research Center. Bouchard's studies improve our understanding of genetic contributions to human disease and exercise performance. He and his colleagues publish an annual review called *Advances in Exercise, Fitness, and Performance Genomics*.

TABLE 1: Targets for Captain America Super Soldier Gene Editing.

PERFORMANCE	GENE/PROTEIN	CLINICAL APPLICATION	HEALTH RISK
INCREASED AEROBIC PERFORMANCE, ENDURANCE	Erythropoietin (EPO)	Improved oxygen delivery in disease (low iron) and trauma	Stroke or heart attack due to increased blood thickness
INJURY RECOVERY	Fibroblast growth factors (FGFs)	Heart disease, wound healing, blood flow	Tumor growth, pathological heart changes
INCREASED WORK ABILITY	Endorphin and enkephalin	Reduced pain sensitivity	Unknown
ENDURANCE CAPACITY AND RESPONSE TO TRAINING	Peroxisome proliferator-activated receptors (PPARs)	Obesity, diabetes, arterial disease	Unknown
OVERCOMING OF INJURIES	SCN9A	Pain disorders	Unknown
SPORTS INJURY RECOVERY	Collagen (COL1A1)	Connective tissue repair	Unknown
MUSCLE SIZE AND STRENGTH	Myostatin inhibitors	Muscle loss in aging	Obesity, diabetes, sarcopenia, weakening of the body
STIMULATION OF BONE AND MUSCLE GROWTH AND RECOVERY	Growth hormone (GH)	Growth deficiency	Disruption in signaling across the body
ENDURANCE AND MUSCLE STRENGTH	TTN	Titin (muscle protein), elasticity in muscle fibers	Unknown
MUSCLE GROWTH, SIZE, AND STRENGTH	Insulin-like growth factor-1 (IGF1)	Prevention of muscle loss in cancer, HIV, and aging	Cancer, heart enlargement

Bouchard suggested to me some of the key gene/protein targets that we would expect to see in someone at the highest physiological level. The list of gene SNPs (single nucleotide polymorphisms) and associated proteins grows longer every year. It also becomes more and more complex, because most of the characteristics of interest are due not to the action of a single gene but more often to the interactions of the products of many genes. Now we have gene targets associated with endurance, strength, response to exercise "trainability," tolerance for exercise, connective tissue repair, and more. Found in Table 1 are some of the potentially important genes/proteins that could be targeted to improve human performance to the ultimate level represented by Captain America.

BUT HOW DO WE DELIVER THE DNA?

Gene doping is the transfer of nucleic acid sequences, normal cells, or genetically modified cells to enhance performance (mainly in a sport context). It's basically the nontherapeutic application of gene therapies developed for diseases or disorders. This means enhancing function in those whose function isn't compromised and revisits a theme we touched upon back in Chapter 2.

Most of the biomedical work on gene therapy has concentrated on inherited diseases and cancer. Since many of these conditions involve dysfunction in cellular metabolism, growth, and viability, it is no surprise that there is growing interest in applying these procedures in situations where the aim is not to forestall disease but rather to dramatically increase human performance. Gene editing is used to change the DNA sequences at specific locations. This is currently done using proteins that cut DNA, like CRISPR-Cas9, Talens, and zinc-finger nucleosis.

If we knew exactly what gene we wanted to transfer into

Steve Rogers, we could produce that gene in a "bacterial plasmid." That's a small strand of DNA that's separated from chromosomal DNA and that can replicate on its own. In our cells, plasmids typically are set up to help with critical functions for survival, such as resistance to environmental toxins. Once we had made enough gene product for our purposes, we'd then purify them and inject them directly into the target tissue. Direct injection is the safest way to get DNA into the target tissue, but it is not as effective as other methods.

A viral vector is more effective. And, yes, that does mean using a virus, one of those things we usually try to avoid, like flu and the common cold. Viruses have evolved to effectively and efficiently penetrate cells, hijack them, and produce their own gene product. In gene therapy, the virus is stripped of all DNA and RNA that it would normally use to replicate or escape from the target cell. The genes of interest are then inserted into the tame virus and injected into the target. A boosting "promoter" allows the inserted gene to be functionally utilized in the target cell and the desired proteins are created.

Ex vivo gene therapy is a technique that's really useful when someone has a serious immunodeficiency. Stem cells are removed from the individual and then a therapeutic gene is introduced to those stem cells. The modified stem cells are then injected back into the patient using the direct DNA technique or viral vectors. This is by far the most successful method for gene therapy in current use. It can be used to insert genes to produce desired proteins, or to modify the expression of a current gene.

COULD WE USE CRISPR TO CUT UP CAP'S GENES?

What if you didn't want to wait for environmental cures to change gene expression? What if, instead, you wanted to do some tinkering on your own, now? Well, then, you'd need to do some gene editing (which presupposes you know what you need to edit in the first place—more on this later). Gene editing still isn't easy, but it was much more difficult before CRISPR came along.

Like so many words in science, CRISPR sounds like something found in a 1950s science fiction movie. In fact, in 2015 *Nature* featured an article by Heidi Ledford called "CRISPR, The Disruptor," a phrase that could easily have been lifted from a Silver Age comic book splash page. And while it sounds artificial and of human construction, CRISPR is derived from the inner workings of the most ancient life on earth—bacteria.

In late 1987, a number of molecular biologists puzzled over the patterns they were seeing in the genome of the bacteria *Escherichia coli*. Later, in 1995, this same sequencing pattern was seen in many other microbes. A certain DNA sequence would be repeated many times but interspersed with unique patterns. This was called "clustered regularly interspaced short palindromic repeats," which was shortened (thankfully) to "CRISPR." So it had a name, but what did it do and why was it there?

Eventually, in 2007, food scientists at Danisco noticed that the unique patterns interspaced with the repeats represented DNA from viruses that attack bacteria. In the same way that your own immune system keeps a representation of pathogens to attack, CRISPR represents an ancient part of the bacterial immune response that allows the bacteria to instantly recognize and target a viral invasion. But CRISPR wouldn't be so superfunctional if all it could do was help recognize an invader. You need to be able to attack it too.

This is where CRISPR-associated proteins—Cas, always found near the CRISPR sequences in the DNA—make their entrance. In 2012, scientists realized that CRISPR could be used in the greatest attack ever—gene editing. Cas go in and chop up the DNA of viral attackers, thus destroying the virus and preventing it from replicating. Science writer Sarah Zhang, at the popular science site io9.com, points to biologist Carl Zimmer and his clear explanation in *Quanta* magazine of how this all works to disable viruses:

> As the CRISPR region fills with virus DNA, it becomes a molecular most-wanted gallery, representing the enemies the microbe has encountered. The microbe can then use this viral DNA to turn Cas enzymes into precision-guided weapons. The microbe copies the genetic material in each spacer into an RNA molecule. Cas enzymes then take up one of the RNA molecules and cradle it. Together, the viral RNA and the Cas enzymes drift through the cell. If they encounter genetic material from a virus that matches the CRISPR RNA, the RNA latches on tightly. The Cas enzymes then chop the DNA in two, preventing the virus from replicating.

The most commonly used Cas enzyme, Cas9, comes from *Streptococcus pyogenes*—the one that gives you strep throat. Cas9 is probably the one you would use to edit your own genetic code. This approach, shown to work in mouse and human cells in 2013, is the most thoroughly studied so far and works basically as follows:

- The CRISPR molecule is programmed to search for a specific nucleotide sequence from the three billion found in the human genome.

- After locating the correct sequence, the CRISPR unwinds the DNA coils and, at a molecular level, snips the sequence out of the strand.

- The DNA strands are then repaired by the body if it's a gene deletion, or a new sequence can be inserted to alter the genome.

If these procedures are performed in an embryonic cell, an egg or a sperm cell, the "edits" will be part of the genetic code that goes to the next generation. As long as you know the right sequence—a guide RNA—to give the Cas9, you can do a cut-and-paste job into your genome.

Say you wanted that myostatin gene deletion we talked about back in Chapter 4. Well, in theory, you would easily be able to use the CRISPR approach to delete it in yourself and maybe get a whole lot stronger. Which is simultaneously thrilling and terrifying. This prospect stimulated one of the mostly hotly debated biomedical ethics debates of 2015. There were calls for a moratorium on the use of CRISPR for editing the human germ line— those cells like sperm and egg cells that pass genetic information to the next generation. Several groups ignored the calls, pressed ahead, and used human embryos as test beds to determine how well the technique might work. And it actually worked. But there were some issues.

There were errors with respect to the targeting associated with the guide RNA. A number of "off-target repeats" had to be eliminated, suggesting that it might be premature to give the technique a clinical application. The errors that occurred could be thought of as the accidental deletions or replacements that turn up in a text document using "replace all" commands. Sometimes, if you don't pay attention, bits of words and phrases get deleted or replaced along with your intended phrases. Except that, unlike a text program with a handy "undo" function, real biology stays changed.

The excitement of CRISPR-Cas9 in real life is its potential

for direct medical use. For example, it might be used to help correct protein expression in Duchenne muscular dystrophy (DMD), an inherited disorder affecting about one in four thousand boys and far fewer girls. Dystrophin is a large protein found in muscle that plays a critical function in muscle contraction. When dystrophin is absent, the muscles slowly change and lose function, reducing the individual's ability to move, and leading eventually to the inability to breathe.

The gene coding for dystrophin is one of the largest in our human genome, with 79 component "exons," and in DMD's many different mutations it can lead to a problem with dystrophin production. Until CRISPR came along, genetic therapies for DMD were limited. This was mostly due to the physically large size of the gene itself, which made it too large to insert using the more common viral vectors. This procedure holds great hope for eventual specific treatment in those with DMD. As Ronald Cohn told *MIT Technology Review* in 2016, "with CRISPR one-of-a-kind treatments are possible and even likely."

An additionally amazing, thrilling, and chilling part of the CRISPR methodology is that you can activate a "gene drive" mechanism that can influence an entire population. The editing doesn't end with the organism you edited. This could have all kinds of problematic implications, but some useful ones as well. It might help us deal with mosquitos, for example, which carry so many nasty pathogens, such as malaria and West Nile virus.

Normally, spreading a genetic mutation in a whole population of organisms takes a long time because a mutation on one of a pair of chromosomes is inherited by only half of the offspring of that organism. A gene drive involves a targeted mutation made using CRISPR, which is then copied to every partner chromosome every generation. This means the mutation spreads extremely rapidly through a whole population. The technique could be used to sweep through and destroy the entire mosquito population carrying malaria or West Nile virus within a single season. Without question, this would constitute a tremendous advance for infectious disease control. But, putting aside for now the predator-prey interactions that would be compromised by wiping out an entire population, molecular biologists are concerned with something else. That something else takes us back to the off-target propensities of the guide RNA. The danger is that the guide RNA itself is likely to mutate as it passes through successive generations and begin to target other areas of the genome, which would then race through the population doing who knows what.

Bioengineer George Church was one of several coauthors who in 2014 warned about the dangers of accidentally releasing gene drives into the wild. They noted that any application must have "high payoff, because it has risks of irreversibility and unintended or hard-to-calculate consequences for other species." Scientists are moving forward on this, despite such dire warnings. As medical geneticist James Wilson (as reported by Heidi Ledford) has said, CRISPR is "ultimately going to have a role in human therapeutics."

This reminds me of a passage from the 1972 science fiction book, *Roadside Picnic*, by Arkady and Boris Strugatsky. They wrote that scientists who know about the current state of research "should be even more scared than the rest of us ordinary folks put together. Because we merely don't understand a thing, but they at least understand how much they don't understand." We await (with trepidation) those well-intentioned, targeted, and regulated applications that will ameliorate the human condition by reducing the burden of disease.

But what about the looming comic book scourge of "mutants"? Would the Super Soldier Serum turn Cap into a mutant?

MUSING ABOUT MUTANTS FROM EXOME TO X-MEN

In September 1963, Marvel Comics billed them as "The Strangest Super-Heroes of All!" "The X-Men" team of "superhuman mutants" was actually invented by Stan Lee and Jack Kirby, but its fictional founder and leader was the science prodigy Professor Charles Francis Xavier. Through his several iterations in the Marvel Universe, Professor X is variously ascribed medical and scientific training but could probably best be described as an expert in medical genetics.

Professor X established a training academy for young mutants as a way to help them employ the powers afforded by their "mutations" and to organize them as the defenders of "ordinary humans" from attacks by their enemies. These included, for example, the clearly labeled "Brotherhood of Evil Mutants."

The theme of "mutants" and mutations permeates all the X-Men comics, graphic novels, and movies and includes more than a morsel of social commentary on what it means to be different. Many readers doubtless identified with the outcasts. But how easily or otherwise can a regular human become a mutant? (Spoiler alert—you are a mutant too.)

Molecular biology, *DNA*, *genes*, and *the genome* are terms that permeate modern culture. The physical bits that we think of as our genes are formed from tiny nucleotide bases that are put together to form chromosomes. In your body you have around three billion pairs of these nucleotides, with around 25,000 to 50,000 genes all arranged on 23 pairs of chromosomes. The collections of those genes in your chromosomes allow you to be the person you are right now with all the physical attributes you possess. Genes can be slightly altered to take on different forms called alleles, and the sets of alleles you have in your genes are what give you your specific genotype.

But the functional role of DNA is to give instructions to the factories in your cells, the ribosomes, for making proteins. Proteins are crucial to life and work in your body to shuttle things within cells, to work as enzymes, provide structural support, and serve as biological motors. They provide your phenotype, the physical expression of your genotype.

Gene expression for a given trait has to do with mutations in the gene pairs. Changes in one allele may lead to expression of a certain phenotype. Humans can adapt to a range of conditions, but many factors beyond genetics have an important influence. There is variation, then, in the extent of responses and adaptations that a person might have, and the genetic influences may not be the dominant factor. That means that the genetic effects will be revealed, expressed, or maximized only in certain specific situations.

Added to the idea of the genome (the map of your genes), we have the proteome (the map of all your proteins) and the exome (the part of your genome that's actually expressed). We still have much to learn about what is normal variation within the human gene pool, but this brings us back to genetic mutations in the biological sense.

Gene mutations are not necessarily either bad or good. It all depends on the broader context. In fact, as Paul Voosen has written, we're all mutants. Many estimates suggest that we humans accumulate 100 to 200 mutations with each generation. In fact, advanced analysis of genomic data has been used to hunt for the clues of "natural selection" in the human genome.

The idea that there could be human mutants with extraordinary powers, such as the healing ability of Wolverine, the chameleon-like shape-shifting of Mystique, and the power of magnetism possessed by Magneto, when taken seriously, strike many of us as pretty cool and worthy of respect. Yet in the plot of the 1980 X-Men comic book, *Days of Future Past*, penned by

Chris Claremont and illustrated by John Byrne, and in the movie based on the same story, mutants and mutations are shown as freaks that frighten broader "human" society. In the comic book we read that in that bleak dystopian year 2013

> there are three classes of people: "H," for baseline human—clean of mutant genes . . . ; "A," for anomalous human—a normal person possessing mutant genetic potential . . . ; "M," for mutant, the bottom of the heap, made pariahs and outcasts by the Mutant Control Act of 1988.

Of course, this fear of mutation in the comics is linked to the fact that genetic mutations in real life often are related to disease states. Cancerous tumors, for example, or conditions such as sickle cell anemia have a pathological basis in red blood cell mutation. But genes aren't just sitting there waiting to give us diseases. As Matt Ridley wrote in *Genome*, "to define genes by the diseases they cause is about as absurd as defining organs of the body by the diseases they get . . . hearts to cause heart attacks and brains to cause strokes."

Evolutionary pressures help to propagate and sustain some genetic mutations even when they are linked to disease states. Sickle cell anemia is an inherited genetic mutation of the oxygen-carrying hemoglobin protein found in red blood cells. The mutation essentially reduces the elasticity of red blood cells, which affects their oxygen-binding function and produces the characteristic curved or "sickle" shape. These red blood cells have a much shorter life cycle than normal red blood cells and can lead to blockages in small vessels. The condition results in a generally shortened life-span.

So if sickle cell anemia is an inherited disorder coming from a mutated allele, why does it persist in the population? Some

MOST ORIGIN STORIES ALLOW FOR NO TRAINING TIME.
HOW DID CAPTAIN AMERICA REALLY GAIN HIS SKILLS?

FIGURE 8: A rare training montage for Steve Rogers en route to becoming Captain America. (From "The Marvels Project" by Ed Brubaker and Steve Epting; © 2011 Marvel Comics Inc.)

scientists have suggested that the persistence of the condition is due to the protective effects of sickle cell anemia in regions where the malaria parasite is present. The malaria *Plasmodium* spends a significant part of its life cycle in red blood cells, and the shortened life cycle of red blood cells in sickle cell anemia interrupts development of the parasite. Consequently, sickle cell anemia has a prophylactic effect against malaria. It's a near-perfect instance of simultaneously "good" and "bad" adaptations. The quotation marks are needed, of course, because the truth is that there is no good or bad biology.

But there are both good and bad in superhero movies. In *X-Men: Days of Future Past*, when evil scientist Dr. Bolivar Trask test-runs his "mutant-o-meter" during the Paris Summit, it not only identifies Mystique as the camouflaged "mutant," it also identifies *all* the humans in the room. What it couldn't do was distinguish with certainty the "good" mutations from the "bad." Whether in the fictional Marvel Universe of Professor X or in our reality, in which medical genetics holds sway, we still have a lot to learn.

We are at the dawn of the age of changing what it really means to be human. Of changing our biology. Changing our genetics, our genome, and our species. We are heading toward the ultimate human.

HOW DID CAP CRAM IN ALL THAT TRAINING?

With all the enhancements in place, our next step is to get Rogers doing a hardcore training program very much like the one I outlined for Bruce Wayne in *Becoming Batman*. After all, Captain America needs to be fit: he's the preeminent fighter in the Marvel Universe. He has engaged in some titanic tussles over the years, including a major league match with Black Panther in the 2016 Marvel Studios movie *Captain America: Civil War*. Unfortunately that film also served up a fight between Captain America and Iron Man that stretched the bounds of credibility.

In any case, "Ultimate Fighter" is what Sam Wilson (aka Falcon) jokingly suggests Steve Rogers consider as a retirement plan in *Captain America: The Winter Soldier*. Okay, that's only going to work if we leave out mystically enhanced folks like Iron Fist and Shang Chi. Here we are talking about enhancing humans with science! Yet training is still needed, and Steve doesn't seem to have done much.

In almost any of the origin stories—take, for example, the 2011 film *Captain America: The Winter Soldier*—Steve Rogers undergoes very little military training. He slings his shield over a considerable distance, but he's also constantly involved in knock-down, drag-out, hand-to-hand fights. He hits, blocks, punches, and kicks with commendable dexterity and power. But how did he get this training? He must have had a lot. In *Avengers* #6 (July 1964), in the story "Meet the "Masters of Evil!" Cap was busy taunting evil scientist Baron Zemo: "I was adept at every form of hand-to-hand combat known to man while you were still safe in your laboratory."

By basing my calculations on the timeline I developed in *Becoming Batman*, I reckon Captain America needs training in martial arts with striking, kicking, and joint locking, as well as projectile weapons, lasting from five to eight years. Quite a bit of training is needed to achieve his level of skill, and even more to fight but not kill. And yet Cap is never shown having more than about a year of training. So how does he do it? The simple answer is this: the motor skills were injected into his brain. That's right—it sounds pretty science fiction, but it's based on some recent science fact.

In 2013, Sam Deadwyler at Wake Forest University and his colleagues at UCLA and the University of Kentucky published a study called "Donor/Recipient Enhancement of Memory in Rat Hippocampus." Now, if you aren't a neuroscientist, this might sound fairly uninteresting. In fact, it's very interesting and very understated—enhancing memory is really important.

Except what the paper really describes is the insertion of memories from one animal that learned something (the donor) into another animal that did not. And then that animal, the recipient, has a new memory.

Deadwyler and co. use approaches related to brain-machine interface and neural prosthetics to show that it's possible to record brain activity from one animal and use that to stimulate the brain of another animal to produce a memory trace. They point out that this could be used to enhance performance, repair damage, or provide training memories in a brain that didn't have any actual training. The ultimate application is to enhance memory in humans and to replace memory in disorders where memory formation and retrieval is failing.

"I think this is a very promising approach, but we are only in the beginning of this journey," explained Mikhail Lebedev, a brain-machine interface researcher at Duke University. "Next demonstrations will be really spectacular. But all the components (to be improved in the future) are already in this one." When I asked him how long it would be before this approach is validated in humans, Lebedev was confident it would happen "within the next decade."

A concept related to this idea of "inserted" training was shown for Bucky Barnes as the Winter Soldier in the Captain America movies. Bucky's character undergoes numerous procedures that "wipe" and then "write" new orders and missions directly into his brain. This is also the only way Cap could get all his skills in such a short period. As Travis Langley and I wrote in the chapter "Training Time Tales with Steve and Anthony" in Langley's 2016 book *Captain America vs. Iron Man: Freedom, Security, Psychology*, superheroes "cannot expect to maintain their skills just sitting around on a Quinjet or at Avengers Tower." We'd also have to instill the enhancements on a reasonable time scale to

FIGURE 9: Brain-machine interface technology could help implant karate kicks in Cap's cortex. (Image courtesy of Kris Pearn)

allow Steve Rogers to adapt to the changes and maintain physiological balance and homeostasis. In the 2011 movie, it all happens in less than a minute; in real life, that would be far too fast to be safe.

Captain America and Iron Man have killed a number of people either as solo characters or as members of various teams. *Captain America Comics* #2 (April 1941) has a story, "Trapped in the Nazi stronghold," in which Cap uses his shield to bash a man in the head and then throws a grenade into a machine gun nest occupied by enemy soldiers. The text says, "the grenade strikes home and the machine gun nest is blown to bits!"—as are, presumably, the soldiers manning the machine gun. This use of lethal force came five issues into Captain America's original run.

But maybe Steve Rogers and Tony Stark really aren't to blame for the number of casualties they've inflicted. To fight with this level of effectiveness, Captain America would need to be at the initial and middle training stages I ascribed to Batman in the timeline referred to above. We know that Cap hasn't trained like Batman. So it's not really the fault of Steve Rogers that he wound up training on the job and using lethal force for a while. And it's even worse for Tony Stark—who needs combat as well as air force training to be a pilot. It's a tough gig, being an Avenger.

Going all the way back to his battlefield debut in World War II, Captain America's brain and body take a massive mauling pretty much every time he is in action. Despite his super soldier enhancements, Captain America is pitched as a human—kind of like us. This means we should think about what would happen to the real human brain behind the giant "A" on his forehead. Is Steve Rogers at risk for concussion and post-concussion syndrome? Probably.

The helmet-ish enhancement on Cap's costume isn't going to do all that much to protect him from injuries to the brain. For sure, it is great protection from fire, and his whole suit protects him from puncture injuries from bullets, other projectiles, and shrapnel. Inside that suit, however, there's a human body that is being subjected to sudden and dramatic starts and stops and brutal impacts. Especially explosive impacts. Impacts like those generate very high angular accelerations that produce concussion.

The word *concussion* is used so often these days that we're in danger of losing sight of its meaning. The Merriam-Webster dictionary says it's "a jarring injury of the brain resulting in disturbance of cerebral function." This is a pretty decent definition. Concussion is one of the most common head injuries found in sports and athletic competitions—and on the battlefield.

Impact to the body can translate into impact on the brain to produce synchronized activity of the neurons of the brain. Many of those cells become briefly and transiently "stunned" (and some killed). Only very slowly do they return to full, normal function. Because of this neuronal disconnect, symptoms of concussion include dizziness, headache, impaired vision, loss of balance and memory, and the inability to concentrate. Although amnesia and loss of consciousness can occur with a concussive impact, Cap doesn't have to be knocked unconscious to have a concussion.

There are also likely genetic factors at play in concussion symptoms. Michael Dretsch and colleagues at the U.S. Army Aeromedical Research Laboratory published a study in 2016 in the *Journal of Neurotrauma* that focused on genetics and risk factors for concussions in active duty soldiers. These folks face the kind of trauma in real life that we imagine a fictional Captain America having to deal with.

Dretsch and his collaborators wanted to find out if there is an association between concussion history and genetic markers that are known to be sensitive to cognitive function. One such marker—and it sounds like it was named by Stan Lee, as it's full-on comic book—is brain-derived neurotrophic factor (BDNF). BDNF is a protein found in the spinal cord and brain. It helps maintain the integrity and function of neurons and their connections. Most of what we know about BDNF in humans is what happens when it's gone. Diseases like Alzheimer's, dementias in aging, and developmental disabilities are associated with low levels of BDNF.

In their study, Dretsch and co. found a strong relationship between a certain polymorphism of the BDNF gene and increased concussion risk. In terms of behavior, this was related to increased aggression and hostility, as determined by personality and psychological testing. Some of these behaviors are likely to worsen with repeated concussion exposures.

Although it might be argued that the regenerative properties of the Super Soldier Serum coursing through Captain America's body might protect him against concussion, I think we'd want to be on the safe side. Cap's environmental exposure to concussive incidents is just too frequent. We'd want to perform a gene edit to make sure his genome does not include the problematic polymorphism. Here, Captain America needs every bit of help he can get.

'OME, 'OME ON THE RANGE

Each concept has its own home, or really they are "-omes." We have the genome (the map of your genes), the proteome (the map of all your proteins), and the variome. The last is the term for the variation within a species as it evolves. And then we have the exome, the part of your genome that's actually expressed. We still have much to learn about what is normal variation within the human gene pool. But we now have a map telling us what is genetically human. A map of the world tells us about the physical layout of our planet. It stays pretty constant unless viewed over millions of years. But now the power to change what "human" means is within reach: it doesn't take millions of years anymore. It can be done within a single animal and a single generation. It can happen fast—possibly too fast.

We are closing in on our DNA with knowledge of what to change and the technology to make the changes. We are getting close to making a real-life Captain America and full-on designer humans. Before tackling deliberate attempts to change those maps, let's think about how they are naturally altered. This brings us to the area of genetic mutations gone wrong.

WHEN SUPERHERO SERUMS GO WRONG, YOU SAY HELLO TO A BIG, GREEN HULK

The backstory of the Hulk has been repeatedly revisited, revised, and reinvented. As with so many Marvel Comics characters, the eventual Avenger "Hulk" was the brainchild of Stan Lee and

Jack Kirby. He made his appearance in *The Incredible Hulk* #1 in May 1962. Lee has said the Hulk was a kind of amalgamation of Robert Louis Stevenson's Dr. Jekyll and Mr. Hyde and Mary Shelley's Frankenstein.

In his original backstory, the Hulk emerges as the alter ego of the mild-mannered physicist Dr. Bruce Banner. Banner is working on a gamma bomb for the U.S. military, but something goes wrong during a test detonation. Instead of viewing the blast from the safety of a lead-lined bunker, Bruce Banner receives a massive dose of gamma radiation. After this, any time that Bruce gets angry or loses control of his emotions, he turns into the raging Hulk.

The Hulk was also one of the founding members of the Avengers. In an early issue, he uttered one of my all-time favorite comic book lines. He is reflecting on his other existence as Bruce Banner and says, "Why shouldn't I be the Hulk?? Why be a puny scientist when I can be the most powerful man walking the earth??!" I ask myself that question just about every day, but I never turn green.

The Incredible Hulk's story trades on both the 1960s paranoia about atomic weapons and the idea of genetic mutations. Somehow the radiation altered something about how Bruce Banner functions. And that something would be his DNA. Of course, it is taken to extremes in the comic book. A variation on the Hulk's origin story was used in the Marvel Ultimates line. In this version, the same army program that successfully created the Super Soldier Serum that produced Captain America in the 1940s tries again. But it fails—again—and instead of a new Cap we have the Hulk.

Captain America's altruism is also wound into many of his stories. In "Crack up on Campus" by Stan Lee in *Captain America* #120 (December 1969), Cap answers an advertisement for a physical education professor at Manning University. While his mission is a success, at the end of the story Cap confesses to S.H.I.E.L.D. commander Nick Fury that he just wasn't up to the task of being a professor.

With these cautions in mind, let's zoom forward and imagine where our species is headed—a super soldier in space and preparing society for the future!

9. BEHOLD THE FUTURE! WHAT'S TO COME FOR CAPTAIN AMERICA?

There is no sensible way in which we must take the
possibility of misuse into account before determining
that something is an enhancement.
—John Harris, *Enhancing Evolution*

The drug may have made me strong, but it did not make me brave!
It may have made me agile, but it did not give me skills! It may have
made me tough, but it did not give me my ideals! Those are all things
I contributed. Those things make up Captain America, not your drug!
—Captain America to Professor Erskine, in *Captain America* #377

What are your real limits as a human being? At some time in our
lives, we've all wondered just what we're truly capable of doing.
We've thought about our limits and how much they constrain us.
Part of that pondering includes wistfully hoping for the limitless
abilities we see in modern pop-culture icons like superheroes,
the mythical heroes of the recent and deep past, and the current

real-life heroes we see in sports. Framing what we'd like to be able to do is a common characteristic of our species. We yearn for more than we have. Our reach perpetually exceeds our grasp.

Many of us have imagined what it might be like to be a superhero. To zoom across the sky like Iron Man, climb like Spider-Man, or have kung fu moves (or maybe Krav Maga—the Israeli Army's self-defense system) like Cap. When we come back to reality, we are forced to recognize the limits on what we can reasonably imagine being.

Ten thousand years ago, *Homo sapiens* experienced a major shift in how we interact with the world. Before that time, hunting and gathering were our main concerns. But then, during the Neolithic Revolution, we began to settle down and plant seeds. This brought about the birth of agriculture. Instead of taking what we could find, we decided to make—or grow—what we needed in order to live. It was the start of a trend. We've been trying to change things around us since before we even thought to use tools. Then, when we got the tools, we made more changes to the environment around us. Now we have the tools to change ourselves directly. We used to have to rely on the slow and haphazard process of evolution. Now we can force evolution to do what we want, when we want it.

When it comes right down to it, we humans have never been satisfied with who we are. Or what we are. There has always been a mismatch between our actual achieved status and what we would like it to be. The flight of Icarus is simultaneously a great story in Greek mythology about effort and achievement and a cautionary tale. Inspiration and hubris are separated by a very fine line.

In the early seventh century, the Hindu scholar-sage Thirumoolar wrote in *Thirumandiram* of the path to immortal divinity. Shiva, one of the main Hindu gods, is a transforming alchemy that creates from a mortal human body an immortal one of solid gold. The idea of alchemy is one of the earliest examples of a deliberate quest to exert our biological human will upon the world around us. It was about trying to change what we see into what we want, even to the extent of changing coal into gold.

The infatuation with alchemy persisted throughout the Middle Ages and into the Renaissance. The concept of the Philosopher's Stone, a metaphysical agent that could transmute matter, was promoted in this period and sustained for centuries after. I argue that this quest continues in our modern-age cult of the body, our unending attempts to change what and who we are. Except that there is now a major difference: we have tremendous powers and abilities at our fingertips. These powers and abilities had no touchstone previously. And as time has gone by, and our abilities have been increasingly amplified, we have also become less patient. We are hurtling forward at a full sprint toward a future we think we can create and control.

The main theme of this book is that all biological beings are immensely pliant and adaptable. This adaptability allows an organism to express the necessary complement of its genome to adapt to the environmental conditions it is exposed to. This is the old nature and nurture interaction expounded by Darwin in his theory of evolution—except evolution used to be a process we had to wait for. Now we may not have to wait. What's more, we may urgently need the new powers we have acquired. The time may be upon us when environmental changes are overtaking our ability to thrive. We may need to thrive in a completely different setting—somewhere off planet earth.

So far, we've had some fun exploring how we could bioengineer a superhero like Captain America. Here I want to consider whether we'll be able to use science and engineering to create a new Jane and John Doe for our imminent future.

SYNTHETIC BIOLOGY, STARDUST, AND THE FANTASTIC FOUR

Let's examine more closely the concept of changing biology. In the words of Pamela Silver and Jeffrey Way in a 2004 article, "Cells by Design," "As the field works to create new living systems that serve a purpose . . . a new foundation for biological understanding should emerge."

Our view of life has to change now that we can change the building blocks of life itself. Usually the term *biology* is interpreted literally as the study of life. While the term first appeared in Western science use in 1791 (in German as *Biologie*), the idea of biology was captured before then in the concept of the "natural sciences." Study of the natural sciences can be traced back to (no big surprise here) at least the ancient Greeks. But until recently the term meant life on earth. And life as we know it now. But biology, like most scientific endeavors, is shaped by the available technology. Biology really took off when the microscope was invented by Dutch anatomist Antony van Leeuwenhoek (1632–1723). Known as the father of microbiology, van Leeuwenhoek was extremely influential in the development of biology as a science. Or, really, the preeminent "life science."

And life science as we know it is built upon knowledge of the plan and function of an organism instructed by the DNA found in the genome. And all those instructions are based on that "simple" molecule of deoxyribonucleic acid. But what if there was an alternative? What if life as we know it isn't the only life we can discover?

We've all done it at one time or another—fixed something with parts from something else. Those parts weren't intended for whatever we were working on, but they "fit," so we used them to replace the part we didn't have, couldn't find, or couldn't make.

This part of our story is about using different parts to form the genetic code. This is synthetic biology.

ENTER XNA AND THE WORLD OF SYNTHETIC BIOLOGY

Synthetic biology brings together approaches and techniques from engineering and life science. Researchers in synthetic biology are focused on creating and constructing new systems and functions that don't exist naturally. Most scientists in this area are trying to understand how life works and the very origins of life itself. Others are trying to figure out how to use the new systems and functions to cure certain diseases.

Scientists have applied their knowledge of synthetic biology to probe inside the 100 trillion cells in the average human body. They have focused on changing the function of cells and working toward ways to communicate between cells. So far, most experiments in synthetic biology involve simple single-celled organisms, such as the yeast *Saccharomyces cerevisiae* (that's our friend used in baking and brewing) and the bacterium *Escherichia coli* (that's the one living happily in our intestines). The cell provides a kind of framework that can be used to engineer changes that include manipulation of the genetic code. This research has led to production of a synthetic rival for DNA, called xeno nucleic acid, or XNA, that can also store and copy genetic information. *Xeno* in this case refers to the alien—literally foreign to the cell—nature of the polymer.

In 2012, Vitor Pinheiro and fellow scientists at the Medical Research Council Laboratory of Molecular Biology in the U.K. created synthetic XNA polymer building blocks by replacing natural sugar components in DNA. Next they evolved polymerases, enzymes that can make XNA from DNA, and others that can do the reverse. This process mimics the normal copying process of RNA and DNA found in natural cells. They discovered that they could create a "heredity" of genetic sequences—the coding of genetic information could be passed along using this artificial process. Then they did something even more interesting. They "stressed" the polymers in test tubes and found that some of them could evolve traits to adapt to the stress. This is the very definition of adaptation for "survival of species" described by Darwin. Their work showed that both the passing on of genetic characteristics and their adaptation of expression could be produced using XNA. For the very first time, this was shown without recourse to RNA and DNA.

This technology could be used now to probe for the origins of life. What other biochemical and biological basis could have been used before we have what we have now? And where might such life exist in the universe? Also, importantly, how can we now specify traits and adaptations in living systems? That brings us to another xeno, but instead of an alien nucleic acid created in a lab, we're talking about the idea of a completely alien biology.

XENOBIOLOGY AND THE BIRTH OF THE FANTASTIC FOUR

Xenobiology is a pretty fantastic word, so why not link it to Marvel's Fantastic Four? It's really a good fit. The Fantastic Four, the brainchild of—who else—Stan Lee and Jack Kirby, were introduced in 1961. They were the first "team-up" comic book that Marvel issued. The original origin story—and I realize that seems like redundant phrasing, but in comics there are many revised origin stories—had Reed Richards, Sue Storm, Johnny Storm, and Ben Grimm gaining superpowers after exposure to cosmic rays while on a space mission. As a result of this dose of cosmic radiation (the actual spectrum is never specified) the powers of stretchiness (Mr. Fantastic, Reed), invisibility (Invisible Woman, Sue), flight and flames (Human Torch, Johnny), and super strength and stone-like skin (The Thing, Ben) are acquired by the team.

I've never really liked this origin story because it relies too much on the "radiation just does stuff" motif. And it doesn't make sense. Not that it has to, but I just like my superheroes better as science fiction than as fantasy. So, please indulge me as I revise the origin story to include the application of real xenobiology. Let's say that the four who become fantastic are infected by an alien retrovirus—a virus that enters the cells and changes the DNA.

A retrovirus is an RNA virus. It gets duplicated in the cell it goes into. Then the host cell uses an enzyme that allows reverse copying (called reverse transcriptase) to create DNA from the RNA. This allows the virus to basically hijack the cell, because the new DNA (from the virus's RNA) is then incorporated into the genome of the cell. This is actually the way gene therapies are applied, so it makes a good illustration of what I mean.

Sure, it's still a bit of a reach to think that the XNA would necessarily be "biocompatible" with human cells. But it's a much smaller reach than cosmic radiation just doing stuff, in my humble opinion. It could also backfire, though, as it does in the story told in *Avengers* #25 in 2012. This story involves the "Super-Adaptoid" that Norman Osborn created to steal genetic characteristics from his enemies, the Avengers. Thus, when he fights the Hulk he gets super strong, when he fights the Vision he gets cool "density control" abilities and so on.

Osborn's plan works until he comes up against Noh-Varr. Noh-Varr, basically a guest member of the Avengers, is a true alien. He is a Kree from across the galaxy and not only does

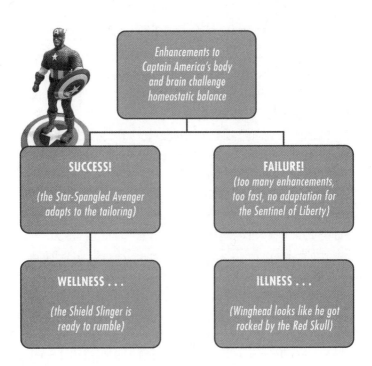

Avengers assemble! Can Steve Rogers survive
the Super Soldier Serum, Vita-Ray treatment,
gene editing, and machine-memory enhancement
required to create the First Avenger?

FIGURE 10: Regulating bodily balance in Steve Rogers.

he have alien DNA—XNA—but he also has access to super-advanced technology. He jumps on Norman Osborn, who then tries to incorporate his superpowers through the "Super-Adaptoid" process. But it all fails because, in the words of Noh-Varr, "my DNA and yours will not mix well. . . . Enjoy trying to adapt to it."

TO INFINITY AND BEYOND

As our species moves forward we will continue to meet challenges—both looming and unforeseen. For us to continue to thrive, the gradual process of evolution that has resulted in our current functional biology designed for life on earth is no longer adequate. In many scenarios, the "survival of the fittest" mantra that many take as the essence of evolution entails too much real and actual death of the less fit.

Some examples of "looming" challenges include virulent pathogens in a global pandemic, rapid climate changes that outstrip our ability to adapt, and life in space or on other planets, such as Mars.

Almost all of our responses to such challenges are just that—responses. To get ahead of the curve we will need proactive planning and implementation, not just crisis management. We need to consider implementing the changes in our species that will enhance our ability to survive and thrive in the future, and not just in the here and now. The science and engineering capacity to make these changes is in our grasp. Implementation takes us further away from the functional biology provided by evolution and toward accelerated evolution to something truly superhuman—a new species adapted by its own hand and optimized for life on earth and beyond.

10. PRE-EVOLVING HUMANITY
FOR FUTURE FRONTIERS
BIOENGINEERED SUPERHEROES IN SPACE

Space: the final frontier . . . to boldly go
where no man has gone before.

—William Shatner as Captain James T. Kirk in the title sequence of *Star Trek*

I think that we've stopped evolving. Because if natural
selection, as proposed by Darwin, is the main mechanism
of evolution . . . then we've stopped natural selection.

—David Attenborough

What humans have accomplished is truly amazing, but it's also
not enough. Not enough for what lies ahead for our species, for
the environmental and health challenges we need to be ready for,
or for our eventual attempts to live in other harsher worlds.

Still, more powerful tools and technologies are within our
reach than ever before. Science and engineering are giving us
ever-stronger real-life abilities. We have used superheroes as
escapist fantasies, but now we are poised to move beyond fantasy

toward an extraordinary future. We may yet become super-heroes. We are certainly on the path to superhuman.

The themes of evolution, genetics, and mutations come up repeatedly in popular culture. These have been particularly high-lighted in comic book series like Marvel's *X-Men*, *Inhumans*, and *Fantastic Four*; in television shows like *Heroes*, Marvel's *Agents of Shield*, and the BBC's *Orphan Black*; and in many science fiction novels where H.G. Wells's *The Island of Doctor Moreau* serves as a thematic starting point. The persistence of these themes, and the variety of media through which they have been given expression, hints at the grip they have in contemporary culture. Our current species, *Homo sapiens sapiens*, has evolved to perform at the highest possible level, given the environment in which we live here on earth. There are myriad ways to deliberately modify and enhance our abilities. Challenges that lie ahead will force us to consider routine biological enhancement. First, however, we really need to make up our minds that we have an obliga-tion to accelerate evolution. This means deliberately altering the nature-nurture balance that underlies adaptive change.

To put it in the simplest terms, nature-nurture interactions arise when environmental signals trigger the expression of the genome. In evolutionary terms, if what's then seen in the proteome is beneficial and adaptive, the organism will flourish and pass that adaptation on to the next generation. And so on. Millennia are required to see meaningful changes in a species. While the environmental cues continue to drive changes in biology, the process continues. Those species that adapt survive, and those that don't won't. This is the so-called "survival of the fittest," long ascribed to Charles Darwin as the centerpiece to the theory of evolution. Yet in our current technological society we mostly have a "soft" environment, in which many of the sculpting pressures are removed.

As a way to gauge how fast meaningful change like this might be achieved, Lauren Koch and colleagues at the University of Michigan Medical School selectively bred rats and tested their functional capacity. Beginning in early 2000, this group began a selective breeding project. They employed a method Gregor Mendel would have found familiar, using Sprague Dawley rats. It's well known that some animals—including humans—respond to exercise training differentially. By breeding rats in pairs based on functional response to exercise training, the researchers created a group of low and high responders. The interesting question was whether a selective breeding program could produce extreme changes in responses in the two groups. The genetic components that affect exercise capacity have an intrinsic component that is found in the untrained state and an extrinsic one that represents the adaptations any animal can produce as a result of training. Basically, the researchers were asking if you could breed rats so that extreme differences in training responses could be realized in later generations.

In 2013, after having selectively bred 15 generations of rats, Koch and his collaborators published their findings. These rats, according to the system they had set up, consisted of "founder" rats (controls), rats that were bred for low exercise response, and those bred for high exercise response. They were trained in treadmill running for eight weeks. The range of responses across the three groups was stunning. The high response rats responded to training by increasing the distance they travelled in a session to approximately 220 meters, while the low response rats actually got worse, their distance declining to about 65 meters. The "control" rats were, on average, in between at about 140 meters. The results demonstrate very clearly that genetic selection can alter functional characteristics in a relatively short time (15 generations).

Yet dramatic shifts in the environment can occur rapidly as

well. The reign of the Sauropods came to an abrupt end 65 million years ago in a so-called extinction event brought about by the earth's collision with a meteor. The resulting dust and ash blocked so much sunlight that the earth was plunged into an ice age. These were not the best conditions for "cold-blooded" animals like dinosaurs. They, and a great many other species, died out because biological evolution could not keep up with the environmental change.

It's reasonable to ask if we face similar environmental pressures now. I said earlier that "soft" environments have become common. Much of the world's population is insulated from the outside (within air-conditioned or heated buildings), typically relies on external forces for movement (cars, public transport), and consumes food that it has not gathered (such as processed food). In effect, we are removed from the normal environmental pressures and triggers that would generate adaptive evolutionary change.

If we take into account the way we live now and the evidence that we are on the edge of periodic global climate change, it's apparent that our species is at a significant disadvantage when it comes to biological adaptation. We're neither culturally nor biologically prepared for the environmental conditions that are bearing down on us. It could be argued that cultural evolution is outstripping our biological function.

Put simply, humans no longer inhabit the environment we evolved in. We don't interact meaningfully with the environment (except, perhaps, electronically), and we are superimposed on a global climate that is rapidly changing. This is a recipe for disaster. On a more positive front, we are clearly headed toward long-term life and colonization of other worlds.

This is where contemporary science and Captain America meet the new frontier of life in space. Space, of course, is not new to Captain America. Cap has been outside earth's atmosphere multiple times, including a legendary (and losing) fight with the "Titanian mutant" supervillain alien Thanos. In 1996's *Infinity Gauntlet* #4 comic, written by Jim Starlin with pencils by George Perez and Ron Lim, Captain America challenges Thanos in battle—showing his dedication and competitive level—but ultimately loses the fight and his shield.

WHAT WOULD LIFE IN SPACE MEAN FOR THE FUTURE OF OUR SPECIES?

Mars has been marked as the next step in the exploration of our solar system. It's not going to be easy: we face challenges with respect to both getting there and surviving once we land. Mars is a harsh place, where we'll have to deal with changes in gravity and the effects of radiation.

Travelers to Mars will experience microgravity in space and about a third of earth's gravity on Mars, changes that raise serious health concerns. NASA has studied the effects of microgravity. Even the physical shape of our cells changes in different gravitational fields. In contortions that would do Plastic Man or the Fantastic Four's "Mr. Fantastic" Reed Richards proud, they move from slightly irregular borders to smoothly rounded ones. Since our cells have biological scaffolding inside them that give and detect changes in shape, along with entry and exit points in the form of ion channels made from proteins dotted throughout, renovations due to changes in gravity lead to alterations in how they function.

Changes in hearing, balance, vision, the distribution of blood in the body (orthostatic intolerance), kidney and urinary function, muscle function, and bone density can also occur. There can be some differences in how changes in these physiological systems are expressed in women and men.

I wanted to find out what real astronauts have experienced. Back in *Inventing Iron Man*, I first explored the "just being off earth and in space" angle by talking with retired NASA astronaut David Wolf. Over his four missions (with both NASA and the Russian Space Agency), Wolf logged a total of 168 days, 8 hours, and 57 minutes in space, including a 128-day "long-duration" mission on the Russian space station Mir. He also spent more than 40 hours "extravehicular"—outside of any spacecraft and just in his spacesuit in space.

Wolf described space as "the harshest environment you can possibly imagine . . . where the temperature can change five hundred degrees in forty-five minutes . . . and there is nobody to help you or to save you. . . . You are completely isolated in a brutal environment. . . . If you make a mistake it could be fatal."

"Just being in space—outside the effects of the gravitational field we've had for all our lives—wreaks havoc on our physiology." Nicole Stott is a retired NASA astronaut who participated on two different space missions, including a long-duration three-month mission on the International Space Station. She's also famous as the first person to paint a watercolor in space and to participate in a tweet-up from space, doing both in October 2009 during NASA's Expedition 21.

When I interviewed Stott for *Project Superhero*, I asked her what it was like when she first returned to earth after three months on the International Space Station. She said that it "felt super heavy . . . like my body felt heavier than I could ever have imagined. I remember coming back and it felt like the lower part of my leg weighed one hundred pounds."

Stott explained that astronauts do a whole battery of tests

and assessments of physiology, and physical and mental function before and after each mission. When she returned from the space station, the scientists "had laid out these small orange safety cones. Like the ones you see on the road, you know? Except these were only about 4 inches tall . . . almost like baby safety cones. And they wanted us to jump over them. . . . It was so hard, and I remember laughing about it and thinking there is no way I can get my body to jump that high."

Just getting up off the floor is hard when you are back on earth after a mission in space. Stott told me that astronauts who were coming home after a long-duration mission would lie down the whole way returning to earth on the space shuttle. Then, "once we were back on land we had to kind of roll off the reclined seats we were on and crawl over to the hatch to get out. There I was, crawling along on my hands and knees trying to get out that hatch. I remember thinking I had to do the heaviest squat lift I had ever done in my life just to lift up my own body to stand up and get out—but just like jumping over those little orange cones, I was able to do it. It's really amazing what our brains and bodies can overcome!"

To get the best possible estimate of what happens to human bodies in space, we'd need to use the time-tested approach of a "twin study." Studies using identical twins are the most rigorous test for a variety of propositions having to do with the effect something has on an individual's inherent genetic predisposition. Typically, twin studies have been used to compare and contrast, for example, talents, skills, or the likelihood of developing a disease or a disorder. Ideally—in a thought experiment, because you couldn't really do this in practice—you would have the twins in isolation for their entire lives, developing independently. NASA and the Russian Space Agency did the next best thing in 2015/2016 as part of a year-long stay on the International Space

Station. The mission commander and NASA astronaut Scott Kelly, who launched in March 27, 2015, on Expedition 43, left his identical twin brother, Mark, back on earth. It was the first time this kind of study had been conducted.

Among the experiments carried out on Mark and Scott was one to examine how ionizing cosmic radiation affects DNA and cellular function. The results have implications for our understanding of aging and the function of the immune system. Unfortunately, the ultimate control was lacking. Ideally, Mark and Scott would have both eaten precisely the same meals and followed the same physical activity patterns. Then the only difference in factors affecting DNA expression would be the environment they were in—space for Scott and Earth for Mark.

Early data from this mission are already showing odd outcomes. Telomere length, a measure of integrity of chromosomes, was temporarily longer in Scott than in his earthbound twin, Mark. (It returned to its prior length on his return to earth.) This appears to contradict the prediction of geneticists that telomere length would be smaller in Scott due to radiation damage. Also, DNA methylation, a cellular mechanism to regulate gene expression, was reduced in Scott but increased in Mark. These observations may represent a type of previously unknown DNA repair mechanism. Much more research and analysis lies ahead, but clearly long-term space flight of even one year has huge effects on human genetics.

The big-picture issue for NASA and other space agencies is the effect that three years in space—roughly the time a there-and-back Mars mission will take—will have on human physiology. To date, 12 months is the longest period that astronauts have remained onboard the International Space Station. Data from Scott Kelly and Russian cosmonaut Mikhail Kornienko (also onboard for 12 months) provide important insights. What we know for sure is that long-duration stays in space have huge effects. Let's pursue our Martian thought experiment a bit further.

The environment in space is literally alien to human physiology—something that has fascinated astrobiologist Dirk Schulze-Makuch at Washington State University. Schulze-Makuch has published extensively on exobiology and extraterrestrial life both in the scientific literature and in writing for the general public—he's the author of four popular books and two novels.

I spoke to Schulze-Makuch in 2016 when he was a guest professor at the Technical University in Berlin, Germany, where he was collaborating on projects related to life in space and on other planets. He was preparing for an upcoming trip to the NASA Jet Propulsion Laboratory in Houston, Texas, where he and his colleagues would present their suggestions for landing and colonizing locations on missions to Mars. We talked extensively about how life in space and on the Red Planet would put extreme demands on life forms—like humans—adapted to conditions on earth. It would be essential to maximize our metabolic efficiency in order to offset—as much as possible—these effects.

Schulze-Makuch believes that if a colony were established on Mars, even taking into account its reduced gravity, we would be able to breed and give birth as we do here on earth. Over generations of life on Mars, we would slowly begin to adapt. Much of that adaptation might be related to the harsh ionizing and ultraviolet radiation that humans would be exposed to. We would gradually, for example, see increased pigmentation—from melanin—in skin exposed to the light on Mars. While that would address the problem of extreme sunburn, it would not necessarily shield us from the effects of ionizing radiation.

And a shield would be needed. Maybe not one made of vibranium, like Cap's, but something as effective. Some microorganisms on earth use minerals like magnesium and manganese to help protect them from radiation. The bacteria *Deinococcus radiodurans*, Schulze-Makuch told me, is one such extremely radiation-resistant organism. This bacterium can survive being frozen, radiated, dehydrated, placed in a vacuum, and exposed to acid. Its superpower is its survivability. It can withstand exposure to more than 5,000 gray (500,000 rad) of ionizing radiation. A dose of this magnitude is known to cause breakage of the DNA strands in other organisms. For reference, exposure to 5 gray would be fatal to a human.

Deinococcus radiodurans is able to survive and resist the effects of radiation because it has rapid DNA repair and multiple copies of its genome. It also has doughnut-shaped DNA strands instead of the double helix we typically find in most organic life on earth. To give you a sense of how much more radiation a body would be exposed to in space, during his 12-month stay on the International Space Station, Scott Kelly was exposed to the same amount of radiation you'd get if you flew between Los Angeles and New York more than five thousand times.

Repairing damage to DNA is a critical function of our cells, and improving our understanding of how it works here on earth helped win Tomas Lindahl, Aziz Sancar, and Paul Modrich—doing independent work—the 2015 Nobel Prize for Chemistry. For life to continue, our cells must divide and create a copy of our DNA in each new cell. If the copy is corrupted, the errors will be handed down to the next generation of cells.

The enzyme DNA helicase helps unwind the double-helical

strand of DNA in your cells, DNA polymerase sticks the new nucleotides to bases on the strands, and then DNA ligase seals everything up and allows your DNA to twist back. The result is replication of perfect copies of identical strands of DNA. Except it doesn't necessarily stay perfect for long. This is biology, after all. DNA strands break down over time. Errors occur due to mistakes in the process of transcription, and, of particular relevance here, damage is inflicted by ultraviolet and other radiation sources.

Tomas Lindahl found an enzyme (uracil-DNA glycosylase) that goes in and can target and cut out damaged nucleotides, and help in repair. Aziz Sancar discovered an enzyme called photolyase. When visible light is shone on DNA it absorbs the light and, using photolyase, repairs damage from UV radiation. Unfortunately, this works well in bacteria but not in humans. However, Sancar later showed that yet another enzyme (excinuclease), found in humans, moves long DNA strands to check for broken bits. When it finds a damaged part of the DNA strand, excinuclease cuts it out and then DNA polymerase comes along and fixes it up.

When the copying process of your DNA is ongoing, sometimes base pairs get mixed up. It doesn't happen very frequently, but with the vast number of cells in your body replicating every 7 to 15 years, a small percentage of errors can cause a lot of problems. Paul Modrich discovered how our cells go in to carry out this "mismatch repair."

Altogether, this awesome work won Modrich, Sancar, and Lindahl recognition from the Nobel committee. And it's all potentially relevant to our consideration of life in space and on Mars. We need to keep working to better understand how the repairs these scientists have described can be optimized by other means, perhaps borrowing from mechanisms in other animals.

Perhaps we could adopt a chimeric approach to creating a

hybrid that includes tardigrade DNA to withstand radiation. Tardigrades are microscopic animals that look like a cross between a bear and a mole. They can withstand just about anything—including the extreme radiation exposure that normally denatures DNA. Takehazu Kunieda and his colleagues at the University of Tokyo discovered that a specific protein in the tardigrade helps in this protective role. It's likely an accidental by-product of the adaptation tardigrades have to withstand severe dehydration. Kunieda and co. manipulated cultured human cells by inserting tardigrade DNA into them. The human cells that had tardigrade DNA—chimeric cells—had a capacity to withstand X-ray damage 40 percent greater than ordinary human cells.

Patrick MacLeod is a physician and medical geneticist living in Victoria, British Columbia. He predicts that in the future, "our understanding and use of DNA will have become incorporated in virtually all aspects of the functional biology of humans and other species." Complete DNA sequencing for everyone could be used in strategies to identify and then prevent the development of disorders and, notably, the "methods for the early diagnosis, treatment and cure of most human diseases. Diagnostic tests will make use of routine point-of-care proteomics [to assess the expression of cellular function]. Monoclonal antibodies and vaccines will be used to treat the patient and methods of RNA interference will be used in cures. Prevention will be the focus of medicine."

MacLeod envisions parallel developments in bioengineering, agricultural science, and adaptation to climate change. These will "ensure a sustainable environment for the production of abundant and appropriate foods, with waste management and abundant clean water for all."

Of course, it's not necessarily all sunshine and roses. Many of these developments could lead to the exploitation of molecular biology in "coercive population control measures, biological weapons, and control of resources by mega-corporations." MacLeod is quick to point out, though, that the most negative outcome he could foresee is that there aren't any outcomes at all. "While the promise is within our current century," MacLeod told me, "a politically imposed funding moratorium can divert resources to last minute and largely futile efforts at mitigating climate change in the hopes of coping with the predicted catastrophic consequences of loss of most of Europe's agricultural lands and major flooding of population centers." I interpret his comments here as a particularly keen observation that useful attempts to functionally adapt our species to deal with climate change may be swamped by unsuccessful attempts to head off climate change itself.

IT MIGHT BE TOO LATE TO EVOLVE ON MARS?

Schulze-Makuch suggested to me that a problem in adapting to the Martian environment might be "conflict between the adaptations needed for the Mars environment and the preexisting physiology we already have." Much of Schulze-Makuch's work has looked at hypothesizing how life could exist on other planets, and he admitted it is a "major challenge to take a highly adapted species like ours and place it out of context on a new planet." He also suggested that over generations of adaptation to life on Mars we would see the "emergence of new species of human."

Mars is our only near-future option as a potential space colony. If we are going to flourish there as a species, we don't have time to wait for evolution to do its work. That's still the case even in a targeted selective breeding project such as the one described earlier. To train new abilities in rats took 15 generations. To achieve comparable changes in humans over 15 generations would take several hundred years. Instead, we may need to consider accelerating human evolution to arrive in a sort of "pre-evolved" state more adapted to the conditions we will face.

In line with the theme of this book, the emergence of the superhuman will be seen in the exploits of the explorers who venture off our world and onto others. This idea has been examined by many science fiction writers, but probably the best example is Frederik Pohl's *Man Plus*, published in 1978. The theme of Pohl's book is that climate and political crises on earth force humans to explore and colonize other planets. Mars is the target, but it's recognized that the planet is profoundly unwelcoming. To deal with this problem, Pohl creates the Exomedicine Project, which has the sole purpose of preparing astronauts to live on Mars. Preparation means changing the form and function of the human body to adapt it to the Mars landscape. Describing one such candidate for the Mars mission, Pohl wrote,

> He did not look human at all. His eyes were glowing, red-faceted globes. His nostrils flared in flesh folds. . . . His skin was artificial. . . . Nothing that could be seen about him was of the appearance he had been born with. . . . All had been replaced or augmented. He had been rebuilt for the single purpose of fitting him to stay alive, without external artificial aids, on the surface of the planet Mars.

This fascinating novel brings out some of the same issues we've addressed here—namely what does it really mean to be human and how far are we willing to go to adapt our biology? Pohl's novel provides an answer: "What it all comes down to is that a colony on the moon can be supported from earth. A colony on Mars cannot. At least a colony of human beings cannot. But what if one reshapes a human being?"

11. THE ETHICAL IMPLICATIONS OF CAPTAIN AMERICA
ARE WE OBLIGED TO ENHANCE OURSELVES?

Human moral status is not assured by our genetic
composition or the arrangement of our cells.

—Insoo Hyun, "Illusory Fears Must Not Stifle Chimaera Research"

You want to protect the world but you don't want it to change.
How is humanity saved if it's not allowed to evolve?

—Ultron to the Avengers, in *Avengers: Age of Ultron*

Time for full disclosure: In writing this book I've arrived at a conclusion far different from the one I envisioned when I started. Where I was once skeptical about enhancing humanity, I now feel strongly that we have an obligation to modify human form and function so we have the best chance to flourish on Earth and in space.

I think that we have to create Captain America. I see our intellect and ability to manipulate all aspects of our environment—including our very biology—as the telling evolutionary adaptation of our species. We must be poised to move from Darwin's "origin of the species" to preservation of the species. Of all species on earth, we are the best placed to do this because we wield the ever-increasing power of science and engineering.

Bioengineering sets the stage for real-life artificial-human brain hybrids and ever more sophisticated technologies to enhance and augment innate function. We can confidently foresee the extension of the concept of brain augmentation to include the "global brain" suggested by Marios Kyriazis of the British Longevity Society. Ultimately this work points to "the new wave of human enhancement."

The development of procedures aimed at brain augmentation should proceed with appropriate caution in neurologically intact "normally functioning" people. The comment made by Rudolf Jaenisch—in the context of the human gene editing controversy—that "we need some principled agreement that we want to enhance humans in this way or we don't" has relevance here. But we need to set the pace ourselves. As Kazuo Ishiguro wrote in the 2005 dystopian novel *Never Let Me Go*,

> When the great breakthroughs in science followed one after the other so rapidly, there wasn't time to take stock, to ask the sensible questions. Suddenly there were all these new possibilities laid out before us, all these ways to cure so many previously incurable conditions. This was what the world noticed the most, wanted the most. . . . But by the time people became concerned . . . it was too late. . . . There was no way to reverse the process. . . . How can you ask the world to put away that cure, to go back to the dark ages? There was no going back.

Since in many cases there really is no going back, thinking about the path ahead is critical.

We've discussed some of the possibilities related to brain augmentation. What other intact cellular interactions in the brain are disrupted by the effect of trans-species implants? What changes in brain structure and function may arise from long-term neuroprosthetic interface? What are the implications for what we now accept as "normal" human behavior and functional capacity? We especially need to establish what societal boundaries—if any—we will place on multispecies transplants and what this means for our species. Many of the related ethical and moral issues are addressed elsewhere in more detail. Going forward, it remains for us as scientists, engineers, and future users of brain augmentation methodologies to proceed with conviction and caution, purpose and care.

The eminent evolutionary biologist and communicator Richard Dawkins, while on *The Daily Show with Jon Stewart* on September 24, 2013, said, "science is the most powerful way to do whatever it is you want to do. If you want to do good, it's the most powerful way of doing good. If you want to do evil, it's the most powerful way to do evil." Never before have we had such influence and so much power.

Future applications of emerging technology can continue to shift us from our subspecies *Homo sapiens sapiens* to the transformative "*Homo sapiens technologicus*"—a species that uses, fuses, and integrates technology to enhance its own function. We move then beyond the limits of function for our species that we originally had back in Chapter 1.

The sports metaphor applies again. Doping in sport is surely an example of enhancing human function beyond the "natural" limitations, but so are many things. In sports, doping creates an uneven playing field—but it also reveals our fascination with unlimited human achievement. Former U.S. Olympic swimmer Shirley Babashoff was 19 years old when she was favored to win several gold medals at the 1976 Summer Olympic Games in Montreal. She came home with just one gold and a handful of silver medals and complained about East German dominance in her events. Later it was shown unequivocally that the East German success was due to their institutionalized steroid regime. Despite that, in the August 8, 2016, issue of *Sports Illustrated*, Babashoff told S.L. Price that she still loves to watch the Olympics—despite the ongoing issue of doping—because "I like to see what people are capable of."

Ruth Slack, professor of cellular and molecular medicine at the University of Ottawa, focuses on neural regeneration and the maintenance of stem cells in the brain. She says, "we hope to repair the damaged brain and prevent the devastating neurological decline found in neurodegenerative diseases." Slack believes advances over the course of the present century will "help us to manipulate the available stem reservoir that will regenerate the adult brain." She is quick to point out that translation of the new technologies to human health is the objective of her work. Yet "before we can effectively and safely activate stem cell populations to repair the damaged brain, a critical understanding of the basic cell biology and physiology of the system we are studying is absolutely fundamental." Slack goes on to talk about where research is headed:

> Motivating and encouraging young researchers to pursue medical research and search for effective treatments of neurodegenerative (and other) diseases is the key to enhance the health and quality of life for future generations to come. In doing this, our contributions to science rest not only on the discoveries made from our own laboratories, but also are multiplied by the brilliant minds that we can motivate to go on and run with the torch.

Our chance of making real progress in accelerating human evolution through science depends in part on our ability to build public support for the task. Communication is key. Slack emphasized this point when she told me, "What we don't contribute enough, is educating the general public—in helping them understand the importance of what we do."

Slack, along with many of the scientists I interviewed, urges caution when it comes to translating new discoveries into therapies. There's great "pressure to 'translate' research as fast as possible," she says, "which is sometimes pushing to therapeutic application before we have sound understanding of the fundamental cellular mechanisms." She uses the example of cellular receptor inhibition for the neurotransmitter glutamate that was rushed too quickly from the bench to clinical trials in stroke treatment. There wasn't enough background, and "clinical trials failed miserably. For many, many years, anytime anyone wanted to study glutamate receptors, people always said 'it will never work,' tarnished by a premature attempt to rush to a clinical trial. This strategy may still be extremely effective, we just need to understand more about the function and regulation of glutamate receptors in the brain."

Whether we arrive at the potential of the vastly modified *Homo sapiens technologicus* depends on many factors. There are many aspects of that transformative evolution that would benefit us greatly, but there are also drawbacks potentially offsetting the impending wins. One thing is certain: we have considerable and growing sway over what our species will become. Toward what future we move can be influenced by our actions.

The words of Alan Kay from 1971 resonate here: "The best way to predict the future is to invent it." Inventing the future is the primary focus of the scientists and engineers at work today. Guiding the implementation of that future is the right and

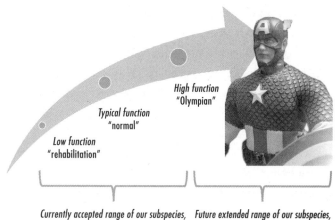

High function
"Olympian"

Typical function
"normal"

Low function
"rehabilitation"

Currently accepted range of our subspecies,
Homo sapiens sapiens

Future extended range of our subspecies,
Homo sapiens technologicus

FIGURE 11: Application of biomedical engineering and science can extend the continuum of human performance abilities toward truly "superhuman" function.

responsibility of us all. It cannot be left solely to the men and women at work in the field and laboratories, nor to those who attempt to regulate their work, the lawmakers and bureaucrats. I believe the future is bright—but there are strings attached. The most important string is that we need input from as many sectors in our society as possible. The decisions that are made will literally affect the future of our species. They cannot be made in isolation.

Science is there to help us, and if the help involves enhancing our abilities, isn't that a good thing? Leigh Montville, writing about the golden age of baseball in the introduction to *The Big Bam: The Life and Times of Babe Ruth*, said this about baseball players: "At the least, these were men who used as much legal modern science as possible to enable them to hit a baseball a long way. They weren't walk-off-the-street human beings."

Our journey together through this book has been fundamentally about advances in science viewed through the lens of popular culture. Many of the issues we've discussed blur the line between science fact and science fiction. That line continues to get hazier with each passing day, as new discoveries and applications of earlier discoveries bring together different approaches from previously unrelated fields.

Science works as a machine of chance effects. Scientists constantly run experiments and test predicted outcomes against a backdrop of random occurrences. Biological evolution is the emergence of chance survival characteristics expanding over millions of years. Any attempts to modify the human germ line—editing sperm or egg cells—have direct implications for the next generation and need to be made carefully. Indeed, many countries have regulations specifically addressing this kind of experimentation. In Canada, for example, the Assisted Human Reproduction Act says that "no person shall knowingly alter the genome of a cell of a human being or in vitro embryo such that the alteration is capable of being transmitted to descendants."

Many of the concerns around biomedical ethics have their origins in concern for their effect on future generations. Steve Rogers knew what he was getting into with the super soldier procedure and consented to participate. But the case of Steve Rogers may be the exception. As pointed out in *Ethics in Health Care: A Canadian Focus* by Eike-Henner Kluge, professor of philosophy at the University of Victoria, "germ line alteration would be performed without the consent of those who are most affected: namely, future generations." Indeed, writer C.S. Lewis in *The Abolition of Man* (published in 1965) suggested that if a society gains power to make descendants "what it pleases, all men who live after it are patients of that power . . . the rule of a few hundreds of men over billions upon billions of men."

Clearly, there is a chance that our genetic enhancements and alterations will go awry. A notable lesson is to be found in the history of the deadly Africanized honey bee, which was created in 1957 when a Brazilian scientist, Warwick Estevam Kerr, interbred European and African honey bees. The result was an extremely aggressive hybrid bee, which was subsequently accidentally released! Yet there are many examples of experiments that have worked out well. Genetically modified crops that can withstand pathogens and harsh environments, for example, have revolutionized agricultural practices around the world.

This is despite the non-scientific arguments and groundless concerns often used to lobby against GMOs or scientifically modified food. For example, Mark Post, professor of physiology at Maastsricht University in the Netherlands, uses muscle stem cells from cows to produce "in vitro" meat for human consumption. Writing in the *Annals of the New York Academy of Sciences* in 2014, Post declared,

> Alternative sources of animal proteins are needed that can be produced efficiently, thereby providing food security with diminished ecological burden. It is feasible to culture beef from bovine skeletal muscle stem cells, but the technology is still under development. The aim is to create a beef mimic with equivalent taste, texture, and appearance and with the same nutritional value as livestock-produced beef.

In a December 24, 2016, commentary from Lois Abraham of the Canadian Press, though, several high-profile celebrity chefs expressed skepticism. British chef Jamie Oliver "doesn't really like the idea of cultured meat." He does allow, though, that "if it's not messing with nature, genetics, and if it's not got [a major] impact on the planet . . . we should probably be open minded."

Chef Anthony Bourdain was more disdainful, saying, "food is supposed to make you happy. . . . It's not nutritional substance. It's not filler. It's not fuel." Well, actually it is all of that stuff and more.

The truth is that in agriculture we have been messing with nature and genetics since forever. We are now just getting more powerful means to do it. And we probably need to continue refining these tools if we want to arrive at the useful future described by Winston Churchill in 1931: "Fifty years hence, we shall escape the absurdity of growing a whole chicken in order to eat the breast or wing by growing these parts separately under a suitable medium."

In any case, although he was not talking about application of science as we now have it, Charles Darwin presaged these ideas in *The Origin of Species* (1859). I can imagine Darwin refuting claims against using the ever-increasing power and precision of science thusly: "Nevertheless so profound is our ignorance, and so high our presumption, that we marvel when we hear of the extinction of an organic being; and as we do not see the cause, we invoke cataclysms to desolate the world, or invent laws on the duration of the forms of life!"

CAN WE REALLY ASSUME WE KNOW ENOUGH TO DO NO HARM?

I do not have the answer to this question. No one does, really. I do, though, think we must go on regardless. Exceeding our limits is a foundational characteristic of our species—now and forever. No matter where we end up in the future, we need to live in the present—a present that is heavily influenced by science and engineering.

Albert Einstein famously said, "It's become appallingly obvious that our technology has exceeded our humanity. The human spirit must prevail over technology." I generally agree with Einstein's words, but I would revise the second part as follows: the human spirit will prevail *through* technology. Milica Radisic at the University of Toronto captured something of what I mean when she told me she is hopeful that, as our species moves forward, "we will be smart enough not to destroy our well-functioning bodies with sedentary life style and poor nutrition, so people who receive the regenerative therapies will only be those who have to receive them for reasons that are beyond their control."

Einstein was worried about controlling technology. But control eventually comes down to life and evolution. "We have to evolve. There's no room for the weak," Ultron said to Quicksilver in the 2015 film *Avengers: Age of Ultron*. Quicksilver asked, "And who decides who's weak?" "Life. Life always decides," was Ultron's reply.

Although many of the ideas we've discussed are based on the ancient concept of exceeding human limitations, most of the science is new. The exact translation of that science will take time. The science we've talked about, like all experiments, requires confirmation by replication: can it be done again? And by refinement: how can it be done better?

We began our journey with a quote from Bobby Kennedy paraphrasing George Bernard Shaw. It's only fitting, then, to close with a revision of another Shaw quote—this one from the 1930 play *Man and Superman*: "The reasonable man adapts himself to the world; the unreasonable one persists in trying to adapt the world. . . . Therefore all progress depends on the unreasonable man."

I suggest a refinement of Shaw's words: all progress depends on the reasonable adaptation of ourselves to the changing world. Once thought unreasonable largely because it seemed impossible, this is the only truly reasonable course available. As Kelly Sue DeConnick said, "impossible is being phased out." On October 28, 2016, at West China Hospital in Chengdu, China, the first test of CRISPR-Cas9 was conducted in humans. The future is now.

Navigating that course is the responsibility of us all. Our society is going to become more reliant on science and engineering and the more everyone understands about science, the better positioned they are to make good, truly informed, decisions. I sincerely hope this book will help ensure that the voices of more people—not just scientists and engineers—are heard when scientific decisions with societal impact are made.

The story of Captain America has always been about societal impact. In the 2010 story "Man out of Time (Part 5)," written by Mark Waid and illustrated by Jorge Molina, Cap echoes this explicitly when he says, "My job is to make tomorrow's world better. Always has been. Once, long ago, I asked Bucky what purpose Captain America served outside of combat. It was a foolish question. There'll always be something to fight for. And I'll always be a soldier."

These qualities were also noted in the 2011 film *Captain America: The First Avenger*. On the eve of the super soldier procedure, Dr. Erskine meets with Steve Rogers to have a philosophical chat. He explains to Steve that earlier attempts at the super soldier procedure were incomplete failures. One of those failures resulted in the creation of the evil nemesis of Captain America, the Red Skull, in the guise of scientist Johan Schmidt. Erskine advises Rogers that his "serum was not ready . . . but more the man [was not ready] . . . the serum amplifies everything. Good becomes great. Bad becomes worse."

FIGURE 12: The inspirational ideals represented by Captain America involve freedom and achievement. (Image courtesy of Andi M. Zehr)

From this conversation, Steve finally learns why he—the 4F reject—was accepted by Dr. Erskine to become Captain America. Steve wonders—why not some stronger soldier? "A strong man," says Erskine, "who has known power all his life may lose respect for that power. But a weak man knows the value of strength. He knows compassion." Presumably also the judicious use of both.

Before leaving Steve for the night, Erskine strikes an imploring note. "Whatever happens," he says, "promise me you will stay who you are . . . a good man." Across his Marvel movie and comic book career, adhering to this promise has been a cornerstone of Steve Rogers's credo. The enduring impact of Captain America's story is the empowerment of that skinny 4F reject into a powerful superhero. For me this resonates with something one of my first martial arts teachers, Shane Higashi, used to tell us about empowerment: "Karate isn't meant to just make the strong stronger; karate makes the weak strong."

While the odyssey of our species continues, many controversies and challenges lie ahead. Our entire society needs to be empowered to engage in the debate and decisions about where we go as a species. Let us, like Captain America, seek a noble path.

Debates about the future of our species will affect the direction taken in scientific research, which in turn affect us all. It's a human problem, and we are all just regular humans.

For now.

AFTERWORD

As I was preparing to write this, I began to think of the subject of E. Paul Zehr's book in a whole new way. As we discuss the idea of superheroes and superhumans, and the path we might take as we leverage technology and medical breakthroughs to improve ourselves, we might also need to ask ourselves if there will be an "afterword" for the human race. Are we on the verge of a huge evolutionary step that we are facilitating for ourselves? Is there a point where "human" just equals "superhuman"? Or will we always just be humans considering ourselves in need of "improvement" and searching for the possibility of a superhuman?

Of course, as Paul conveys so well, there is always the underlying concept of the parallel between who we consider to be superheroes and our idea of a superhuman. What is a superhero? Dictionary definitions such as "a benevolent fictional character with superhuman powers" and "an exceptionally skillful or successful person" actually work quite well because they cover how we look at the characters we have come to know through fiction, but also address the reality of how we look at ourselves and the "real" people around us and in our lives.

My guess is that if you randomly ask someone, "What is a superhero?" you will likely get a response that includes one of the well-known fictional characters like Captain America or Wonder Woman, and maybe even a statement like "someone who fights evil." But if you were to have a conversation about the idea of superheroes, it would likely lead to the second definition, where you'd consider your "real-life heroes" to be like your parents or someone doing something exceptional that you find impossible to imagine for yourself. In one way or another, all of these super-heroes are also superhuman to us. Because of this broad defini-tion, though, there are many different ways an afterword to Paul's book could be approached. I have chosen to approach it through my own personal perspective as an astronaut.

In 2000—a year known for its threat of millennium bugs and crashing computers, and a year when those of my generation had been promised flying cars and the Jetsons and when Arthur C. Clarke's vision from *2001: A Space Odyssey* was only a year away—I was incredibly fortunate to be selected into the 18th class of NASA astronauts.

The title *astronaut* has always been associated with *hero*. As one of these people myself, that's a little difficult to wrap my mind around, because while I definitely consider the astronauts before me as heroes, I don't think of myself that way. What humans have done collectively to prepare astronauts to live and work for extended periods of time in space, and to develop the technology to transport and sustain us in space, is pretty amazing. Just 50 years ago we were wondering if humans could function and survive at all in the microgravity environment of space, and now we have demonstrated how adaptable and capable we are in such an extreme environment. The human ability to adapt and learn and develop and utilize the technology necessary to make it all happen is pretty super in itself.

Fifty years ago, for the first time, we as humans experienced our planet as a whole from space. *Earthrise*, the iconic image taken by Apollo astronaut Bill Anders, represents the first time that humans in space had seen this life-changing view with their own eyes and captured it for the rest of us. Through this quest to venture off our planet and into space we have given ourselves a whole new vantage point for appreciating the planet we all share and for better understanding our place in this grander universe.

Through this journey in space, we also are learning more and more about the risks the space environment poses to us as humans. Our planet, our spaceship Earth, although it exists in space, is a beautifully designed place for us to survive, situated at the perfect distance from the sun and blanketed by our fragile, thin, and yet amazingly protective atmosphere. The microgravity, or what is more commonly called the "zero g," environment that we have been living and working in so far in the space outside our planet's atmosphere poses some very interesting opportunities and challenges for us as human beings. As a result, when I think of the "superhuman" aspects of space flight, two things come to mind—what our bodies are capable of in the microgravity environment that they aren't capable of here on Earth, and the countermeasures that we have to employ to simply survive in space and later return safely to Earth.

As an astronaut, and more importantly as a mom, I am a true believer in the value of human space flight and exploration. I understand the importance of these missions and know that everything we are doing through our exploration and habitation in space is ultimately about improving life here on Earth. But I also know that I couldn't wait to get to space so that I could experience some of the more unusual and just plain fun aspects of being there. Who hasn't dreamed for their whole life of how awesome it would be to have the strength of Superman and to fly on your own like he does? Who hasn't longed for the athleticism, strength, and strategic skills of Wonder Woman? And, like them, to use these superhuman traits and the technology around you for the good of humanity?

To float, to fly, in space is a very surreal thing, but interestingly it is something that your brain and your body very quickly adapt to. Moving naturally and gracefully through a three-dimensional space. No sense and no real consideration for what's "up" or "down." Effortlessly moving yourself and anything else around you. Shifting your concern from how to get moving to how you'll stop yourself once you are moving. It's the most liberating and comfortable feeling—all with the opportunity to float in front of a window presenting the most stunning view of our home planet below you. The ability to do these things does make you feel superhuman.

Every fictional superhero has their strengths, like all of us, but they all have their weaknesses too—their "Kryptonite." The downside of our bodies and our brains adapting so quickly to a new environment like space is that they figure out very quickly how to really adapt. I know that sounds a bit counterintuitive, because isn't "adapting" exactly what we want our bodies to do? Just one example of our Kryptonite, an example of where this adaptation doesn't work to our advantage, is that our bodies figure out very quickly that you don't need bones or muscles to survive in zero gravity, and so they don't waste any more energy maintaining them. While this transformation might be OK if you were planning to live the rest of your life in zero gravity, it certainly won't support a healthy or safe return to Earth. As a result, astronauts currently employ a number of different "countermeasures" to counteract these effects on our bodies. With respect to bones and muscles, our exercise protocols are the number one counter-measure to their loss. We have a mix of resistive and aerobic exercise equipment that we use two hours per day to counteract these effects and to maintain our bone density and muscle mass. These exercise protocols have proven extremely effective as a counter-measure to the space environment, and have also demonstrated their effective application to similar terrestrial health issues associated with bone density and muscle mass loss.

As we continue on this human journey of space exploration, farther from the protection of our planet, we must also continue to understand the risks associated with the "human" aspects of that exploration. Our current work on the International Space Station and our history of living and working in space give us a very good baseline understanding, but we also realize there are a lot of things that we just don't know we don't know yet. Just like 50 years ago when we were first sending humans into space, some of the things we thought would be the most challenging weren't and some of the things we thought would be insignificant also weren't—proving that we have to be as well prepared as we can be, but we also have to be ready for the surprises.

Over the history of humanity here on Earth and in space, we have continually employed technology and new understanding of how our bodies work to improve our quality of life. Every one of us very likely has some example of this augmentation that makes us "better" than we would be without it—simple and complex things like clothing, eyeglasses, hearing aids, pace-makers, hip replacements, organ transplants, prosthetics, medi-cations of any type. These are tools we've chosen to improve, but in the simplest sense they also make us superhuman, giving us superpowers that we wouldn't have on our own.

Into our future we will also continue to leverage technology and our better understanding of how our bodies operate in order to improve our quality of life both on and off our planet. We are only just beginning to discover what those new tools might be and developing an understanding of what their longer-term implications might be, but like our fictional superheroes we go forward with the hope that there is always an approach of good versus evil. The first corrective/magnifying eyeglass, called a "reading stone," was used over a thousand years ago, and eyeglasses more like our own today were initially used in the 13th century. For a long time, the use of eyeglasses was considered a sign of weakness, but it's very clear now that the technology behind these pieces of glass or plastic is one of the most accepted and important inventions in human history. There are, and I'm certain there will continue to be, many more examples of the evolution of technology and its use for improving the quality of human life.

In closing, I'd like to say that I think the term *superpower* is relative, as demonstrated by the contrast between how our bodies work in the zero-gravity environment of space, where we all have the ability to fly, and what we experience when we come back from space and wonder if we can even carry the load of our own body off the spaceship. I would also like to come back to the idea of the "afterword" for us humans. I believe that we are blessed with the gifts of curiosity and creativity and invention, which we will continue to use for the benefit of humanity. I believe that we can look at every stage of our evolution in comparison to the past and see how we have improved our quality of life. I believe we are super-human already, but that won't stop us from continuing to search for even more "super" human possibilities.

NICOLE STOTT, Artist and NASA Astronaut (retired)

ACKNOWLEDGMENTS

I remain deeply influenced by my two main scientific mentors, Dr. Digby Sale at McMaster University and Dr. Richard Stein at the University of Alberta. These professors spurred my interest in physiology and neuroscience and also a sense of belonging to the broader community and doing work with societal impact. I point out the ongoing support of folks from the Comic Arts Conference: Drs. Peter Coogan (co-founder), Kathleen McClancy, and Randy Duncan. Especially the influence of Drs. Travis Langley and Mark D. White.

I conducted many interviews during the writing of *Chasing Captain America* and I am grateful to those who agreed to speak and correspond with me: Nicole Stott, Kevin Strange, and Drs. Dan Ferris, Milica Radisic, Peter Reiner, Thomas Stieglitz, Molly Shoichet, Jon Wolpaw, Doug Weber, Mikhail Lebedev, John Morley, George Church, Dirk Schulze-Makuch, Patrick MacLeod, Ruth Slack, and Claude Bouchard.

Thanks to Warren Ellis, Daniel H. Wilson, and Mark D. White for advance blurbs on this book. Thank you to Simon Whitfield for his gracious foreword and to Nicole Stott for her thoughtful afterword.

I thank the many test readers—Drs. Keir Pearson, Leigh Anne Swayne, Trevor Barss, and Taryn Klarner, and Yao Sun, Greg Pearcey, Hilary Cullen, Steve Noble, and Robert Frost—of both the very early drafts (sorry about that) and the final version (less sorry about that). Special thanks to Marj Wilder for providing comments and supporting various aspects of production.

I remain impressed by the level of energy and professionalism of my publisher, ECW Press, and thank Crissy Calhoun, Jack David, David Caron, Rachel Ironstone and my fantastic editor Jonathan Webb. The helpful work of my copy editor Merrie-Ellen Wilcox is gratefully acknowledged! Thank you to my agents Sam Hiyate and Ali McDonald at the Rights Factory. Thanks to the University of Victoria for its support of community engagement and knowledge sharing through entertaining outreach efforts like this. UVic truly does embrace its motto: *multitudo sapientium sanitas orbis*—A multitude of the wise is the health of the world.

Thank you, readers of my prior books and blogs. The positive feedback you provide in reviews, letters, and emails continues to inspire me every day.

Above all, to my family, thank you for your steadfast support.

BIBLIOGRAPHY

COMICS AND GRAPHIC NOVELS

The Amazing Spider-Man #1 (March 1963), Stan Lee (writer) and Steve Ditko (art)

The Avengers #1 (September 1963), Stan Lee (writer) and Jack Kirby (pencils)

The Avengers #4 (March 1964), Stan Lee (writer) and Jack Kirby (pencils)

The Avengers #5 (May 1964), Stan Lee (writer) and Jack Kirby (pencils)

The Avengers #6 (July 1964), Stan Lee (writer) and Jack Kirby (pencils)

Avengers #25 (2012), Jonathan Hickman (writer) and Mike Deodato (pencils)

Captain America Comics #1 (March 1941), Joe Simon and Jack Kirby (co-writer, co-pencils)

Captain America Comics #3 (May 1941), Joe Simon and Jack Kirby (co-writer, co-pencils)

Captain America Comics #20 (July 1943), Stan Lee (writer) and Al Avison (pencils)

Captain America Comics #28 (July 1943), Syd Shores (pencils)

Captain America #78 (September 1954), John Romita (art)

Captain America #100 (April 1968), Stan Lee (writer) and Jack Kirby (pencils)

Captain America #109 (January 1969), Stan Lee (writer) and Jack Kirby (pencils)

Captain America #120 (December 1969), Stan Lee (writer) and Jack Kirby (pencils)

Captain America #377 (September 1990), Mark Gruenwald (writer) and Ron Lim (pencils)

Captain America #445 (November 1995), Mark Waid (writer) and Ron Garner (pencils)

Captain America #446 (December 1995), Mark Waid (writer) and Ron Garner (pencils)

Captain America #447 (January 1996), Mark Waid (writer) and Ron Garner (pencils)

Captain America #448 (February 1996), Mark Waid (writer) and Ron Garner (pencils)

Captain America: Man out of Time, Part 5 (March 2011), Mark Waid (writer) and Jorge Molina (pencils)

Captain Marvel #12 (June 2013), Christopher W. Sebela and Kelly Sue DeConnick (writers) and Filipe Daniel Moreno De Andrade (pencils)

Fantastic Four #1 (November 1961), Stan Lee (writer) and Jack Kirby (pencils)

The Incredible Hulk #1 (May 1962), Stan Lee (writer) and Jack Kirby (pencils)

The Incredible Hulk #181 (November 1974), Len Wein (writer)
and Herb Trimpe (pencils)

Infinity Gauntlet #4 (1996), Jim Starlin (writer) and George
Perez and Ron Lim (pencils)

The Marvels Project (2011), Ed Brubaker (writer) and Steve
Epting (pencils)

The Secret Defenders #7 (September 1992), Roy Thomas
(writer) and Andre Coates (pencils)

Tales to Astonish #27 (January 1962), Stan Lee and Larry
Lieber (writers) and Jack Kirby (pencils)

Ultimate Hulk vs. Iron Man: Ultimate Human (graphic novel,
2008), Warren Ellis (writer) and Cary Nord (pencils)

The Ultimates: Ultimate Collection (graphic novel, 2010), Mark
Millar (writer) and Bryan Hitch (pencils)

The Uncanny X-Men #1 (September 1963), Stan Lee (writer)
and Jack Kirby (pencils)

The Uncanny X-Men #141 (January 1981), Chris Claremont
and John Byrne (writers) and John Byrne (pencils)

Wolverine: Origin (graphic novel, 2009), Paul Jenkins and Bill
Jemas (writers) and Joe Quesada and Andy Kubert (pencils)

MOVIES AND TELEVISION PROGRAMS

Ant-Man (2015, Marvel Studios)

Avengers: Age of Ultron (2015, Marvel Studios)

Captain America: Civil War (2016, Marvel Studios)

Captain America: The First Avenger (2011, Marvel Studios)

Captain America: The Winter Soldier (2014, Marvel Studios)

Dawn of the Living Dead (1968, Image Ten [as an Image Ten
Production], Laurel Group, Market Square Productions,
Off Color Films)

Doctor Who (1963–1989, 2005–present; BBC TV)

Guardians of the Galaxy (2014, Marvel Studios)

The Incredible Hulk (2008, Marvel Studios)

Limitless (2011)

Marvel's The Avengers (2012, Marvel Studios)

The Six-Million Dollar Man (1974–1978; ABC TV)

The Wolverine (2013, Fox Studios)

X-Men 3: The Last Stand (2006, Fox Studios)

X-Men: Days of Future Past (2014, Fox Studios)

X-Men Origins: Wolverine (2009, Fox Studios)

BOOKS, JOURNALS, NEWSPAPERS, MAGAZINES, AND WEBSITES

Ahmad, C.S., W.J. Grantham, and R.M. Greiwe. "Public Perceptions of Tommy John Surgery." *Physician and Sportsmedicine*, 40, no. 2 (2012): 64–72.

Asimov, I. *I, Robot*. New York: Doubleday, 1950.

Balague, G. *Messi*. London: Orion, 2013.

Bamford, S.A., R. Hogri, A. Giovannucci, A.H. Taub, I. Herreros, P.F.M.J. Verschure, P. Del Giudice, *et al*. "A VLSI Field-Programmable Mixed-Signal Array to Perform Neural Signal Processing and Neural Modeling in a Prosthetic System." *IEEE Transactions on Neural Systems and Rehabilitation Engineering* 20 (2012): 455–467. doi:10.1109/TNSRE.2012.2187933.

BBC. "In Quotes: Joan Rivers." September 5, 2014. http://www.bbc.com/news/entertainment-arts-29075239.

Beatty, S.C., A. Dowgill, and A. Dougall. *The Avengers: The Ultimate Guide to Earth's Mightiest Heroes*. London: DK Publishing, 2012.

Becker, A.J., C.E. McCulloch, and J.E. Till. "Cytological

Demonstration of the Clonal Nature of Spleen Colonies
Derived from Transplanted Mouse Marrow Cells." *Nature*
197 (1963): 452–4.

Bhullar, B.-A.S., Z.S. Morris, E.M. Sefton, A. Tok, M. Tokita,
B. Namkoong, A. Abzhanov, *et al*. "A Molecular Mechanism
for the Origin of a Key Evolutionary Innovation, the Bird
Beak and Palate, Revealed by an Integrative Approach
to Major Transitions in Vertebrate History." *Evolution* 69
(2015): 1665–77. doi:10.1111/evo.12684.

Bryant, H. *Juicing the Game: Drugs, Power, and the Fight for the
Soul of Major League Baseball*. New York: Viking, 2005.

Buck, J. *Lucky Bastard: My Life, My Dad, and the Things I'm
Not Allowed to Say on TV*. New York: Dutton, 2016.

Burkett, B. "Technology in Paralympic Sport: Performance
Enhancement or Essential for Performance?" *British
Journal of Sports Medicine* 44 (2010): 215–20. doi:10.1136/
bjsm.2009.067249.

Burns, L. "'You Are Our Only Hope': Trading Metaphorical
'Magic Bullets' for Stem Cell 'Superheroes'." *Theoretical
Medicine and Bioethics* 30 (2009): 427–42. doi:10.1007/
s11017-009-9126-0.

Camporesi, S. "Oscar Pistorius, Enhancement and Post-
Humans." *Journal of Medical Ethics* 34 (2008): 639.
doi:10.1136/jme.2008.026674.

Church, G.R., and E. Regis. *Regenesis: How Synthetic Biology
Will Reinvent Nature and Ourselves*. New York: Basic Books,
2012.

Clark, V.P., and R. Parasuraman. "Neuroenhancement:
Enhancing Brain and Mind in Health and in Disease."
NeuroImage 85 Part 3 (2014): 889–94. doi:http://dx.doi
.org/10.1016/j.neuroimage.2013.08.071.

Costanzo, J.P., M.C.F. do Amaral, A.J. Rosendale, and R.E.

Lee. "Hibernation Physiology, Freezing Adaptation and Extreme Freeze Tolerance in a Northern Population of the Wood Frog." *Journal of Experimental Biology* 216 (2013): 3461–73. doi:10.1242/jeb.089342.

Cyranoski, D. "CRISPR Gene-Editing Tested in a Person for the First Time: The Move by Chinese Scientists Could Spark a Biomedical Duel between China and the United States." *Nature* 539 no. 479 (2016). doi:10.1038/nature.2016.20988.

Cyranoski, D. "Ethics of Embryo Editing Divides Scientists." *Nature* 519 no. 7543 (2015): 272. doi:10.1038/519272a.

Darwin, C. *The Origin of Species by Means of Natural Selection.* London, 1859.

Davies, J. "Program Good Ethics into Artificial Intelligence." *Nature* 538 no. 7625 (October 20, 2016): 1.

Davila, A.F., and D. Schulze-Makuch. "The Last Possible Outposts for Life on Mars." *Astrobiology* 16 no. 2 (2016): 159–68. doi:10.1089/ast.2015.1380.

Deadwyler, S.A., R.E. Hampson, A. Sweat, D. Song, R.H.M. Chan, I. Opris, I., T.W. Berger, *et al.* "Donor/Recipient Enhancement of Memory in Rat Hippocampus." *Frontiers in Systems Neuroscience* 7 (2013). doi:10.3389/fnsys.2013.00120.

Deary, I., W. Johnson, and L.M. Houlihan. "Genetic Foundations of Human Intelligence." *Human Genetics* 126 no. 1 (2009): 215–32. doi:10.1007/s00439-009-0655-4.

Desrivieres, S., A. Lourdusamy, C. Tao, R. Toro, T. Jia, E. Loth, G. Schumann, *et al.* "Single Nucleotide Polymorphism in the Neuroplastin Locus Associates with Cortical Thickness and Intellectual Ability in Adolescents." *Molecular Psychiatry* 20 no. 2 (2015): 263–74. doi:10.1038/mp.2013.197.

di Lampedusa, Tomasi, G. *The Leopard*. New York: TIME, 1960.

Di Nardo, P., D. Singla, and R.-K Li. "The Challenges of Stem Cell Therapy." *Canadian Journal of Physiology and Pharmacology* 90 no. 3 (2012): 273–4. doi:10.1139/y2012-016.

Domengos, P. *The Master Algorithm: How the Quest for the Ultimate Learning Machine Will Remake Our World*. Philadelphia: Basic Books, 2015.

Dretsch, M.N., N. Silverberg, A.J. Gardner, W.J. Panenka, T. Emmerich, G. Crynen, G.L. Iverson, *et al*. "Genetics and Other Risk Factors for Past Concussions in Active-Duty Soldiers." *Journal of Neurotrauma* 34 no. 4 (2017). doi:10.1089/neu.2016.4480.

Egner, I.M., J.C. Bruusgaard, E. Eftestøl, and K Gundersen. "A Cellular Memory Mechanism Aids Overload Hypertrophy in Muscle Long after an Episodic Exposure to Anabolic Steroids." *Journal of Physiology* 591 no. 24 (2013). doi:10.1113/jphysiol.2013.264457.

Eiraku, M., N. Takata, H. Ishibashi, M. Kawada, E. Sakakura, S. Okuda, Y. Sasai, *et al*. "Self-Organizing Optic-Cup Morphogenesis in Three-Dimensional Culture." *Nature* 472 no. 7341 (2011): 51–56.

Epstein, S.L. "Wanted: Collaborative Intelligence." *Artificial Intelligence* 221 (2015): 36–45. http://dx.doi.org/10.1016/j.artint.2014.12.006.

ESPN. "Tommy John Surgery." December 5, 2012. http://espn.go.com/mlb/topics/_/page/tommy-john-surgery.

Forget, T. *The Creation of Captain America*. New York: Rosen, 2016.

Franco, A. "Pat Mendes, Top American Weightlifter, Banned for HGH Use." *USA Today* (April 16, 2012). http://usatoday30.usatoday.com/sports/olympics/

story/2012-04-16/weightlifter-pat-mendes-suspended
-hgh/54326514/1.

Gehrz, C. "Olympic Records over Time." (Blog post, July 27, 2012.) https://pietistschoolman.com/2012/07/27/olympic-records-over-time/.

Gibson, D.G., J.I. Glass, C. Lartigue, V.N. Noskov, R.Y. Chuang, M.A. Algire, J.C. Venter, *et al.* "Creation of a Bacterial Cell Controlled by a Chemically Synthesized Genome." *Science 329* no. 5987 (2010): 52–56. doi:10.1126/science.1190719.

Graff, V. "David Attenborough: 'We're Lucky to Be Living When We Are, Because Things Are Going to Get Worse'." *RadioTimes* (September 20, 2013). http://www.radiotimes.com/news/2013-09-20/david-attenborough-were-lucky-to-be-living-when-we-are-because-things-are-going-to-get-worse.

Grau, C., R. Ginhoux, A. Riera, T.L. Nguyen, H. Chauvat, M. Berg, G. Ruffini, *et al.* "Conscious Brain-to-Brain Communication in Humans Using Non-Invasive Technologies." *PLoS ONE* 9 no. 8 (2014): e105225. doi:10.1371/journal.pone.0105225.

Grierson, B. *What Makes Olga Run?: The Mystery of the 90-Something Track Star and What She Can Teach Us About Living Longer, Happier Lives.* New York: Henry Holt, 2014.

Han, X., M. Chen, F. Wang, M. Windrem, S. Wang, S. Shanz, M. Nedergaard, *et al.* "Forebrain Engraftment by Human Glial Progenitor Cells Enhances Synaptic Plasticity and Learning in Adult Mice." *Cell Stem Cell* 12 (2013): 342–53. doi:http://dx.doi.org/10.1016/j.stem.2012.12.015.

Harris, J. *Enhancing Evolution: The Ethical Case for Making Better People.* Princeton, NJ: Princeton University Press, 2007.

Hashimoto, T., D.D. Horikawa, Y. Saito, H. Kuwahara, H. Kozuka-Hata, T. Shin-I, T. Kunieda, *et al.* "Extremotolerant Tardigrade Genome and Improved Radiotolerance of Human Cultured Cells by Tardigrade-Unique Protein." *Nature Communications 7* (2016): 12808. doi:10.1038/ncomms12808.

Herreros, I., A. Giovannucci, A.H. Taub, R. Hogri, A. Magal, S. Bamford, P.F. M. J. Verschure, *et al.* "A Cerebellar Neuroprosthetic System: Computational Architecture and in vivo Test." *Frontiers in Bioengineering and Biotechnology* 2 (2014): 14. doi:10.3389/fbioe.2014.00014.

HuffingtonPost. "Thomas Beatie, the 'Pregnant Man,' Wants a Fourth Child." September 5, 2012. http://www.huffington post.com/2012/09/04/thomas-beatie-pregnant-man-fourth-child_n_1855318.html.

Hutchison, C.A., R.Y. Chuang, V.N. Noskov, N. Assad-Garcia, T.J. Deerinck, M.H. Ellisman, J.C. Venter, J.C. *et al.* "Design and Synthesis of a Minimal Bacterial Genome." *Science* 351 no. 6280 (2016). doi:10.1126/science.aad6253.

Hyun, I. "Illusory Fears Must Not Stifle Chimaera Research." *Nature* 537 no. 7620 (2016): 1.

Ishiguro, K. *Never Let Me Go.* New York: Random House, 2005.

Jaenisch, R., and B. Mintz. "Simian Virus 40 DNA Sequences in DNA of Healthy Adult Mice Derived from Preimplantation Blastocysts Injected with Viral DNA." *Proceedings of the National Academy of Sciences of the United States* 71 (1974): 1250–4.

Kaku, M. *Physics of the Future: How Science Will Shape Human Destiny and Our Daily Lives by the Year 2100.* New York: Anchor, 2011.

Kluge, E.-H. *Ethics in Health Care: A Canadian Focus.* Toronto: Pearson Education Canada, 2012.

Koch, L.G., G.E. Pollott, and S.L. Britton. "Selectively Bred Rat Model System for Low and High Response to Exercise Training." *Physiological Genomics* 45 (2013): 606–14. doi:10.1152/physiolgenomics.00021.2013.

Kota, J., C.R. Handy, A.M. Haidet, C.L. Montgomery, A. Eagle, L.R. Rodino-Klapac, B.K. Kaspar, *et al.* "Follistatin Gene Delivery Enhances Muscle Growth and Strength in Nonhuman Primates." *Science Translational Medicine* 1 no. 6(2009): 6ra15. doi:10.1126/scitranslmed.3000112.

Kramer, M. "Elon Musk: Artificial Intelligence Is Humanity's 'Biggest Existential Threat'." *LiveScience* (October 27, 2014). http://www.livescience.com/48481-elon-musk-artificial-intelligence-threat.html.

Kyle, U.G., and C. Pichard. "The Dutch Famine of 1944–1945: A Pathophysiological Model of Long-Term Consequences of Wasting Disease." *Current Opinion in Clinical Nutrition and Metabolic Care* 9 (2006): 388–94. doi:10.1097/01.mco.0000232898.74415.42.

Kyriazis, M. "Systems Neuroscience in Focus: From the Human Brain to the Global Brain?" *Frontiers in Systems Neuroscience* 9 (2015). doi:10.3389/fnsys.2015.00007.

Langley, T., and E.P. Zehr. "Training Time Tales with Steve and Anthony: The I/O Psychology of Getting Better at Being Super." *Captain America vs. Iron Man: Freedom, Security, Psychology*, ed. T. Langley, 65–76. New York: Sterling Books, 2016.

Lebedev, M.A., and M.A. Nicolelis. "Chapter 3: Toward a Whole-Body Neuroprosthetic." *Progress in Brain Research* 194 (2011): 47–60).

Ledford, H. "CRISPR: Gene Editing Is Just the Beginning." *Nature* 531 (2016): 4.

Lewis, C.S. *The Abolition of Man*. London: Oxford University Press, 1943.

Libbrecht, R., and L. Keller. "The Making of Eusociality: Insights from Two Bumblebee Genomes." *Genome Biology* 16 no. 1 (2015): 75. doi:10.1186/s13059-015-0635-z.

Looso, M., J. Preussner, K. Sousounis, M. Bruckskotten, C.S. Michel, E. Lignelli, T. Braun, *et al.* "A de novo Assembly of the Newt Transcriptome Combined with Proteomic Validation Identifies New Protein Families Expressed During Tissue Regeneration." *Genome Biology* 14 no. 2 (2013): R16. doi:10.1186/gb-2013-14-2-r16.

Lynch, G., L.C. Palmer, and C.M. Gall. "The Likelihood of Cognitive Enhancement." *Pharmacology Biochemistry and Behavior* 99 no. 2 (2011): 116–29. doi:http://dx.doi.org/10.1016/j.pbb.2010.12.024.

McDonell, T. "In My Tribe." *Sports Illustrated* (2011, November 28). http://www.si.com/vault/2011/11/28/106135818/in-my-tribe.

Melrose, K. "After 23 Surgeries to Look Like Superman, 'Man of Plastic' Meets His Kryptonite." CBR.com, February 2, 2015. http://www.cbr.com/after-23-surgeries-to-look-like-superman-man-of-plastic-meets-his-kryptonite/.

Mnih, V., K. Kavukcuoglu, D. Silver, A.A. Rusu, J. Veness, M.G. Bellemare, D. Hassabis, *et al.* "Human-Level Control through Deep Reinforcement Learning." *Nature* 518 no. 7540 (2015): 529–533. doi:10.1038/nature14236.

Montville, L. *The Big Bam: The Life and Times of Babe Ruth*. New York: Doubleday, 2006.

Morley, J.E.C., and S. Colberg. *The Science of Staying Young*. New York: McGraw Hill, 2007.

Nacu, E., E. Gromberg, C.R. Oliveira, D. Drechsel, and E.M. Tanaka, E.M. "FGF8 and SHH Substitute for

Anterior–Posterior Tissue Interactions to Induce Limb Regeneration." *Nature* 533 no. 7603 (2016): 407–10. doi:10.1038/nature17972.

O'Doherty, J.E., M.A. Lebedev, P.J. Ifft, K.Z. Zhuang, S. Shokur, H. Bleuler, and M.A.L. Nicolelis. "Active Tactile Exploration Using a Brain-Machine-Brain Interface." *Nature* 479 no. 7372 (2011): 228–31. http://www.nature .com/nature/journal/v479/n7372/abs/nature10489 .html#supplementary-information.

Pais-Vieira, M., M. Lebedev, C. Kunicki, J. Wang, and M.A.L. Nicolelis. "A Brain-to-Brain Interface for Real-Time Sharing of Sensorimotor Information." *Scientific Reports* 3 no. 1319 (2013). http://www.nature.com/ srep/2013/130228/srep01319/abs/srep01319.html#supple mentary-information

Parsons, P. *The Science of Doctor Who*. Baltimore: Johns Hopkins University Press, 2010.

Petruzzo, P., S. Testelin, J. Kanitakis, L. Badet, B. Lengelé, J.-P. Girbon, J.-M. Dubernard, *et al*. "First Human Face Transplantation: 5 Years Outcomes." *Transplantation* 93 no. 2 (2012): 236–40. doi:10.1097/TP.0b013e31823d4af6.

"Pictures of First Person to Undergo Plastic Surgery Released." *Telegraph*. August 28, 2008. http://www .telegraph.co.uk/news/uknews/2636507/Pictures-of-first -person-to-undergo-plastic-surgery-released.html.

Pinheiro, V.B., A.I. Taylor, C. Cozens, M. Abramov, M. Renders, S. Zhang, P. Holliger, *et al*. "Synthetic Genetic Polymers Capable of Heredity and Evolution." *Science* 336 no. 6079 (2012): 341–4. doi:10.1126/science.1217622.

Pohl, F. *Man Plus*. New York: Orb Books, 1976.

Portin, P. "The Birth and Development of the DNA Theory

of Inheritance: Sixty Years Since the Discovery of the Structure of DNA." *Genetics* 93 no. 1 (2014): 302.

Reardon, S. "Performance Boost Paves Way for 'Brain Doping'." *Nature 531* (2017): 2.

Regalado, A. "Can CRISPR Save Ben Dupree?" *MIT Technology Review* (October 17, 2016). https://www.technologyreview.com/s/602491/can-crispr-save-ben-dupree/?set=602660.

Ridley, M. *Genome: The Autobiography of a Species in 23 Chapters*. New York: Harper Collins, 1999.

Sainato, M. "Stephen Hawking, Elon Musk, and Bill Gates Warn About Artificial Intelligence." *Observer* (August 19, 2015). http://observer.com/2015/08/stephen-hawking-elon-musk-and-bill-gates-warn-about-artificial-intelligence/.

Sarzynski, M.A., R.J.F. Loos, A. Lucia, L. Pérusse, S.M. Roth, B. Wolfarth, C. Bouchard *et al.* "Advances in Exercise, Fitness, and Performance Genomics in 2015." *Medicine and Science in Sports and Exercise* 48 (2016): 1906–16. doi:10.1249/mss.0000000000000982.

Schuelke, M., K.R. Wagner, L.E. Stolz, C. Hubner, T. Riebel, W. Komen, S.J. Lee, *et al.* "Myostatin Mutation Associated with Gross Muscle Hypertrophy in a Child." *New England Journal of Medicine* 350 (2004): 2682–88.

Sebastiani, P., N. Solovieff, A.T. DeWan, K.M. Walsh, A. Puca, S.W. Hartley, T.T. Perls, *et al.* "Genetic Signatures of Exceptional Longevity in Humans." *PLoS ONE* 7 no. 1 (2012): e29848. doi:10.1371/journal.pone.0029848.

Shelley, M. *Frankenstein; or, The Modern Prometheus*. London, 1818.

Silver, P., and J. Way. "Cells by Design." *Scientist* (September 27, 2004).

Smith, C. "High School Basketball Player's Stunning Wingspan." (Blog post, June 17, 2011.) http://sports.yahoo.com/highschool/blog/prep_rally/post/top-junior-hoops-prospect-may-have-longest-wingspan-ever?urn=highschool,wp2931.

Staples, A. "Think Big: 6'9", 396-pound Daniel Faalele Has Coaches Drooling—and He's Never Played a Down." *Sports Illustrated* (March 6, 2017). http://www.si.com/college-football/2017/03/07/daniel-faalele-img-academy-recruiting.

Strugatsky, A., and B. Strugatsky. *Roadside Picnic*. Chicago: Chicago Review Press, 1972.

Tan, T.C.J., R. Rahman, F. Jaber-Hijazi, D.A. Felix, C. Chen, E.J. Louis, and A. Aboobaker. "Telomere Maintenance and Telomerase Activity Are Differentially Regulated in Asexual and Sexual Worms." *Proceedings of the National Academy of Sciences of the United States of America* 109 (2012): 4209–14. doi:10.1073/pnas.1118885109.

Tanghe, K.B. "A Historical Taxonomy of Origin of Species Problems and Its Relevance to the Historiography of Evolutionary Thought." *Journal of the History of Biology* (2016): 1–61. doi:10.1007/s10739-016-9453-8.

Voosen, P. "We Are All Mutants: On the Hunt for Disease Genes, Researchers Uncover Humanity's Vast Diversity." *Chronical of Higher Education* (March 24, 2014). http://www.chronicle.com/article/We-Are-All-Mutants/145393.

Wallace, D.C. "Bioenergetics in Human Evolution and Disease: Implications for the Origins of Biological Complexity and the Missing Genetic Variation of Common Diseases." *Philosophical Transactions of the Royal Society B: Biological Sciences* 368 no. 1622 (2013). doi:10.1098/rstb.2012.0267.

White, M.D. *The Virtues of Captain America*. West Sussex, UK:
    Wiley-Blackwell, 2014.

Wilkins, J. "Defining Evolution." National Center for Science
    Education, 2001. https://ncse.com/library-resource/
    defining-evolution-0.

Wilson, D.H. *Amped*. New York: Doubleday, 2012.

Wilson, D.H. "Dude, Where's My Jetpack? A Look at the
    Future That Refuses to Arrive." *Discover* (February 26,
    2007). http://discovermagazine.com/2007/feb/jetpack-
    future-technologies.

Windrem, M. S., S.J. Schanz, C. Morrow, J. Munir, D.
    Chandler-Militello, S. Wang, and S.A. Goldman. "A
    Competitive Advantage by Neonatally Engrafted Human
    Glial Progenitors Yields Mice Whose Brains Are Chimeric
    for Human Glia." *Journal of Neuroscience* 34 (2014): 16153–
    61. doi:10.1523/jneurosci.1510-14.2014.

Witze, A. "Astronaut Twin Study Hints at Stress of Space
    Travel: Unusual Study of NASA's Scott and Mark
    Kelly Finds Gene-Expression Shifts During Nearly a
    Year in Space." *Nature* (January 26, 2017). doi:10.1038/
    nature.2017.21380.

Zehr, E.P. *Becoming Batman: The Possibility of a Superhero*.
    Baltimore: Johns Hopkins University Press, 2008.

Zehr, E.P. "Future Think: Cautiously Optimistic About Brain
    Augmentation Using Tissue Engineering and Machine
    Interface." *Frontiers in Systems Neuroscience* 9 (2015).
    doi:10.3389/fnsys.2015.00072.

Zehr, E.P. *Inventing Iron Man: The Possibility of a Human
    Machine*. Baltimore: Johns Hopkins University Press, 2011.

Zehr, E.P. *Project Superhero*. Toronto: ECW Press, 2014.

Zehr, E.P. "The Potential Transformation of Our Species by

Neural Enhancement." *Journal of Motor Behavior* 47 no. 1 (2015): 73–8. doi:10.1080/00222895.2014.916652.

Zhang, S. "Everything You Need to Know about CRISPR, the New Tool That Edits DNA." *Gizmodo* (June 5, 2015). http://gizmodo.com/everything-you-need-to-know-about-crispr-the-new-tool-1702114381.z

INDEX

*Underlined page numbers indicate tables;
numbers in italics indicate figures.*

A

Aboobaker, Aziz, 110

Abzhanou, Arhat, 58

activin A, 67

adaptability

    Captain America as repre-
        senting, *17*

    genome and, 147

    hormones and, 49

    of human body, 13

    of human brain, 82

    to stress, 24

adenine, 53

adenosine triphosphate (ATP),
    106–7

Africanized honey bees, 175

aging. *see* biological aging

agriculture

    birth of, 146

    genetics and, 175–76

alchemy, 146–47

alleles, 54, 111, 132–33

ALS (Lou Gehrig's disease), 84–85

amino acids, 55

AMPA receptor, 92

ampakines, 92

Amped, 37–38

anabolic steroids, 49–50, 73, 116

androstenedione, 49–50

Ant-Man, 113–14

Ant-Man (movie), 114–15

Anya, BeeJay, 17–18

apolipoprotein E precursor
(APOE), 80

apoptosis, 108–9

arm, reconstructive surgery for,
46–47

artificial intelligence (AI)
conflict with biological beings,
98
evolution of, 97–98
existential threat of, 99–100
need for ethical codes for, 102

Asimov, Isaac, 101

Assisted Human Reproduction Act
(Canada), 174

assistive technologies
ethical implications of, 34–35
loss of privacy and, 25
misuse of, 36

astrocytes, 82

astronauts, 159–61, 181–86

athletes
as cultural icons, 14
extreme performance of, 18
homeostasis and, 48
steroid use in, 48–50, 73
as superheroes, 23–24

Austin, Steve (*Six Million Dollar
Man*), 74–75

The Avengers, 113–14, 143

The Avengers (comic book)
#1, 12–13
#4, 6–7, 13, 16
#6, 137
#25, 151–52

The Avengers (film), 12, 62, 111

Avengers: Age of Ultron, 97–98, 177

Avery, Oswald, 52

axolotl (*Ambystoma mexicanum*),
105

B

Babashoff, Shirley, 170–71

bacterial plasmid, 126

Bamford, Simeon, 88

Banner, Bruce, 143

Barnes, Bucky
bionic arm of, 29–30
freezing of, 111–12
inserted training and, 138–39

Batman, 121

*Becoming Batman: The Possibility of
a Superhero*, 14, 62, 136

Berger, Theodore, 88

Bhullar, Bhart-Anjan, 58

biocompatibility, 151

bioengineering
brain augmentation and,
169–70
brain–machine interface and,
87–88
possibilities of, 14–15, 122–123
risks of, 41
superhuman function and, 173

biological aging
process of, 108–9
reversal of, 118–19

biological determinism, 19

biology, 148

Biolux Research, 106–7

biomedical ethics, 174

Blackburn, Elizabeth H., 109

bone marrow transplantation, 64

Bouchard, Claude, 123–24

bovine muscular hypertrophy, 67

brain

augmentation of, 169–70

cell transplantation research
    on, 82–84

disconnect between anatomy
    and physiology of, 75–76

enhancing function in, 77–78,
    88

NSI-189 and, 92–96

stimulating function of, 32–33,
    96–97

tissue engineering in, 81–84

brain-derived neurotrophic factor
    (BDNF), 80, 141

brain–machine interfaces (BMIs)

control of, 84–86

inserted training and, 138–139

interfaces for movement, 90

memory enhancement and, 138

research into, 36–37

sensory interfaces and, 90

brain-to-brain interfaces, 89

Braun, Thomas, 105–6

Burkett, Brendan, 31–32

Burns, Lawrence, 61, 63

C

calcium, as modulator of neuron
    function, 82

Calment, Jeanne Louise, 117

Camporesi, Silvia, 34

cancer

epigenetic changes and, 72

rapamycin (mTOR) and, 115

war on, 117–18

candidate genes, 79–80

Captain America

altruism of, 143–44

creation of, 5, 10–12

ideals represented by, *179*

muscle strength of, 60–61

origin story of, 6–7, 12, 137

revival of, 16–17

as virtue personified, 28

see also Rogers, Steve

Captain America Comics

#1, 5, 12, 29

#2, 139

#3, 6

#28, 61

#78, 12

#100, 7

#109, 7, *8*

#120, 143–44

#444–54, 7

super soldier procedure in, *6,
    8, 40*

Captain America: The First Avenger,
    7, 11–12, 60, 178

*Captain America: The Winter
    Soldier*, 7–9, 12, 87–88, 111–13,
    121, 137

carbon blades, 30–31

Cas9, 128

see also CRISPR-Cas9

catabolic hormones, 49

cellular memory, 70–73

cellular regeneration, 106

cerebral cortex, 80

chalones, 67

Chavez, Herbert, 45

chimeras, 81–82, 84

chimeric cells, 165

chimeric-embryo research, 82–84

chromosomes

    mutations and, 130

    structure of, 53–54, 132

    telomeres and, 109, 161

Church, George, 117–19, 131

Clark, Vincent, 33–34

clean competition, 48

climate change, 153, 157, 165–66

clock speed, 82

cognitive enhancement, 96–97

cognitive enhancers, 91

collaborative intelligence, 100

collagen (COL1A1), 124

comic book superheroes, birth of, 5

complex phenotypes, 110–11

compression of morbidity, 115

concussion

    brain-derived neurotrophic
        factor (BDNF) and, 141

    defined, 140

    genetic factors for, 141

    symptoms of, 140

cortisol, 49

cosmetic surgery, 44–45

Costanzo, Jon, 112–13

Court of Arbitration for Sport, 31

Crick, Francis, 52

CRISPR-associated proteins (Cas),
    126–27

CRISPR-Cas9

    calls for moratorium on, 129

    first human test of, 178

    overview of, 125–31

cryoprotection, 112–13

cybernetics, 34

cyborgs

    Dan Ferris on, 35–36

    REX (robotic exoskeleton), 35

cytochrome C oxidase, 106

cytosine, 53

D

Darwin, Charles, 22–23, 176

Davies, Jim, 102

Dawkins, Richard, 170

Days of Future Past, 133–34

Deadwyler, Sam, 137–38

Deary, Ian, 79–80

deoxyribonucleic acid. *see* DNA
    (deoxyribonucleic acid)

Desrivières, Sylvane, 80–81

Di Nardo, Paolo, 66

dino-chickens, 58–59

disease-risk alleles, 111

DNA (deoxyribonucleic acid)

    coding within, *53*

    delivery of, 125–26

    expression of, 19

    function of, 55–56, 133

    mismatch repair of, 164

    repairing damage to, 163–64

    as software of the cell, 122–23

    structure of, 52–54

    see also synthetic DNA

DNA helicase, 163–64

DNA ligase, 164

DNA methylation, 161

DNA polymerase, 164

DNA sequencing, 165

Domingos, Pedro, 99
doping. *see* drug doping;
    gene doping
Dr. Who, 34–35
Dretsch, Michael, 141
drug doping
    in baseball, 48–50
    to enhance abilities, 122
    gene doping and, 18
Duchenne muscular dystrophy
    (DMD), 130
dystrophin, 130

E

Egner, Ingrid, 70–73
Einstein, Albert, 177
Eiraku, Mototsugu, 65
electroencephalography (EEG)
    as control signal, 84–85
    detecting signals with, 89
endorphin, 124
enhancement
    acceptable vs unacceptable, 32
    ethical implications of, 32, 34
    in sports, 29–30
enkephalin, 124
environment
    effect on genetic material of,
        21–23
    soft environment, 155, 157
epigenetics, 22, 70–71
epigenome, 71
Epstein, Susan, 100
erythropoietin, 124
ethical issues
    around assistive technologies,
        34–35

    around enhancement, 32
    Simon Whitfield on, 3
eusociality, 116
Evans, Chris, 7
evolution
    acceleration of, 155, 166–67,
        172
    chance effects and, 174
    defined, 23
    as theme in popular culture, 155
    as too gradual, 153
ex vivo gene therapy, 126
excinuclease, 164
exercise response, 156
exome, 133, 142
Exomedicine Project, 167
exons, 130
exoskeletons, 25, 35–37, 117

F

famine, effect on genetic material
    of, 21–22
Fantastic Four, 150–51
Ferris, Dan, 35–36, 87
fibroblast growth factors (FGFs),
    58, 106, 124
follistatin, 69
Franklin, Rosalind, 52
Fritsch, Gustav, 32

G

Gates, Bill, 99–100
Gehrz, Chris, 23–24
gene doping, 18, 122–23, 125
gene drive mechanism, 130–31
gene editing
    CRISPR-Cas9 and, 126–31

gene editing, *continued*
    infectious disease control and, 131
    targets for, 124, 125
gene expression
    mutations and, 57, 133
    regulation of, 71
gene therapies, 125–26, 151
genetically modified organisms
    (GMOs), 175
genetic amplification, 79
genetic engineering, 51–52
genetic mutations, 133–36
genetics
    Charles Darwin and, 22–23
    overview of, 51–59
    and risk factors for concussion,
      141
    as theme in popular culture, 155
genome, 51, 133, 142
genome-wide complex trait
    analysis, 78–79
genotypes, 51, 54
germ line, modification of, 174
global brain, 169
glucose, 112–13
glutamate, 92, 172
GMOs (genetically modified
    organisms), 175
Google DeepMind, 98
Google Glass, 38
Grau, Charles, 89
gravity, 158, 184, 186
Greider, Carol W., 109
growth factors
    effect of, 67
    genetic manipulation of, 69

growth hormone (GH), 39–41, 49,
    70, 124
guanine, 53
guide RNA, 129, 131

H
hacking into body parts, 37, 87
Han, Xiaoning, 82–83
Hawking, Stephen, 100
Hayflick, Leonard, 114
Hayflick limit, 114
healing ability, 103–4
heroes
    American heroes, 28
    classical heroes, 13, 28
    see also superheroes
Herreros, Ivan, 88
hippocampus, 92
Hitzig, Eduard, 32
homeostasis, 48
Homo sapiens
    evolution of, 17, 155
    Neolithic Revolution and, 146
    new subspecies of, 170
Homo sapiens technologicus, 170, 172
Hongerwinter, 21–22
hormones
    cortisol, 49
    growth hormone (GH), 39–41,
      49, 70, 124
    muscles and, 116–17
    overview of, 47–48
    testosterone, 69–70
Hulk
    backstory of, 142–43
    see also The Incredible Hulk

human ability, normal limits of,
    26, 29
human ambition, 28
human genome, 56
Human Genome Project, 56–57
human identity, loss of, 90
human performance abilities
    continuum of, 33
    genes associated with, 69
human potential
    genetics and, 71
    limits of, 145–46
    Simon Whitfield on, 1–2
Hyun, Insoo, 84

I
Icarus, 28
immunosuppression, 69
in vitro meat, 175
The Incredible Hulk
    #1, 143
    #181, 104
infectious disease control, 131
inserted training, 138–139
insulin-like growth factor 1 (IGF1),
    124
insulin-like growth factor 2 receptor
    (IGF2R), 80
intelligence
    as complex trait, 78
    genetic influence on, 78–80
International Association of
    Athletic Federations (IAAF),
    30–31
International Space Station, 158–59,
    184

*Inventing Iron Man: The Possibility of
    a Human Machine*, 14, 36–37, 85
invulnerability, 103
ionizing cosmic radiation, 161–62
Iron Man, 62
Ishiguro, Kazuo, 169

J
Jaenisch, Rudolf, 19, 169
Jobe, Frank, 46
John, Tommy, 46

K
Kay, Alan, 172–73
Keller, Laurent, 116
Kelly, Mark, 161
Kelly, Scott, 161, 163
Kerr, Warwick Estevan, 175
Kirby, Jack, 5, 114, 150
Kluge, Eike-Henner, 174
Koch, Lauren, 156
Kornienko, Mikhail, 161
Kota, Janaiah, 68
Kunieda, Takehazu, 165
Kyriazis, Marios, 169

L
Lebedev, Michael, 89–90, 138
Lee, Stan, 6, 114, 150
Lernaean Hydra, 81
Lieber, Larry, 114
life in space, 153
life science, 148
life-span, 109–10, 115
Lindahl, Tomas, 164
Li, Ren-Ke, 66

longevity, 110–11, 116
long-term potentiation (LTP), 83
Lou Gehrig's disease (ALS), 84–85
Lynch, Gary, 91

M
machine-brain-machine interfaces, 101
MacLeod, Patrick, 165
Major League Baseball, 48–50
malaria, 135–36
Mars, 158, 162–64, 166–67
Marvel Comics, 5–6
McCulloch, Ernest, 64
McGwire, Mark, 49–50
memory
    enhancement of, 88, 137–38
    long-term potentiation (LTP) and, 83
Mendel, Gregor, 51–52
messenger RNA, 55–56
Messi, Lionel, 39–40
metabolome, 51, 71
Methuselah gene, 110–11
microgravity, 158, 182–83
Miescher, Friedrich, 52
mind-reading devices, 90
Mintz, Beatrice, 19
missing heritability, 79
mitochondria, 106, 117
Mnih, Volodymyr, 98
Modafinil (Provigil), 91–92
Modrich, Paul, 164
Morley, John, 116–17
mouse-human chimera, 82–84
muscles
    Captain America and, 60–61
    cellular memory in, 72–73
    hormones and, 116–17
    loss of with aging, 116–17
    myostatin in, 67–68
    stem cell applications in, 65
Musk, Elon, 99
mutant healing factor, 105
mutants, 131–32, 134
mutations
    gene expression and, 57
    of myostatin and activin A, 67
    as theme in popular culture, 155
myostatin, 67–69
myostatin inhibitors, 124

N
Nacu, Eugeniu, 106
NASA, 159–61
National Core for Neuroethics, 96
natural human ability, 26, 29
natural selection, 22–23, 133
nature and nurture, 21–22
Neolithic Revolution, 146
nervous system, direct interfaces with, 86–87
neural enhancement, 32–33
neural interface technologies, 86, 101–2
neural networks, playing video games, 98
neural prosthetics, 138
neurodegenerative diseases, 117, 171
neurogenesis, 92
neuromodulation, 91
neuroplastin (NPTN), 80–81
neuroprosthetics, 36, 85, 88, 170
newts, 105

Nicolelis, Miguel, 37

Noh-Varr, 151–52

non-selective augmentation, 25–26

nootropics, 91

normal human ability, 26, 29

NSI-189

    neurogenesis and, 92–96

    recreational use of, 93–96

    side effects of, 94–95

nucleotides, pairing of, 52–53, 55–56, 71

nurture

    effect on genetic material of, 21–22

    see also nature and nurture

O

off-target repeats, 129, 131

On the Origin of Species by Means of Natural Selection, 22–23

Operation Rebirth, 7

OrthoPulse, 107

OsseoPulse Bone Regeneration System, 106

Össur Flex-Foot Cheetah, 30

oxidative damage, 117

oxytocin receptor (OXTR), 80

P

Pais-Vieira, Miguel, 89

pandemics, 153

Parasuraman, Raja, 33–34

peroxisome proliferator-activated receptors (PPARs), 124

phenotypes

    defined, 51

    expression of, 54, 57

proteins and, 133

Philosopher's Stone, 147

photobiomodulation, 106–7

photolyase, 164

Pinheiro, Vitor, 149–50

Pistorius, Oscar, 30–32

planarian worms, 105, 110

plastic surgery, 44–45

Pohl, Frederik, 167

polypeptides, 55

Post, Mark, 175

post-concussion syndrome, 140

pre-evolution, 154–55, 166–67

privacy

    loss of, 25

    violation of, 87

Professor X, 77, 132

programmed cell death, 108–9

prosthetics

    restoration vs enhancement of ability with, 29–30

    in sports, 30–32

    see also neuroprosthetics

protein expression, 59

proteins, function of, *53*, 55–56

proteome, 51, 71, 133, 142

proteomics, 165

Pym, Hank, 113–14

Pym Particles, 114, 116

R

radiation, 158, 161–62

radiation-resistant organisms, 163

Radisic, Milica, 24–26, 177

rapamycin (mTOR), 115

rapid DNA repair, 163

rapid self-healing
    photobiomodulation and,
        106–7
    possibilities of, 105
recessive traits, 54
reconstructive surgery
    cosmetic surgery, 44–45
    history of, 44
    ulnar ligament reconstruction
        surgery, 46
regenerative medicine
    future of, 24, 38
    misuse of, 39
    repairing heart cells, 25
Regenesis; How Synthetic Biology
    Will Reinvent Nature and
    Ourselves, 119
Regis, Ed, 119
Reiner, Peter, 96
Reinstein, Professor, 5–7, 11
repetitive transcranial magnetic
    stimulation (rTMS), 33–34, 89
retroviruses, 151
reverse transcripterase, 151
REX (robotic exoskeleton), 35
ribonucleic acid. *see* RNA
    (ribonucleic acid)
ribosomes, 55
Ritalin (methylphenidate), 91
RNA (ribonucleic acid), 19, *53*, 56
    see also specific types, e.g. guide
        RNA
RNA viruses, 151
robotic assistants, 117
Rogers, Steve
    medical assessment of, 121
    origin story of, 5–9

qualities of, 11
regulating body balance in, *152*
training of, *135*–137
see also Captain America
Russian Space Agency, 159–60

S
Sancar, Aziz, 164
sarcopenia, 116–17
Sasai, Yoshiki, 65–66
Schuelke, Markus, 67
Schulze-Makuch, Dirk, 162–63, 166
science fact vs science fiction, 174
SCN9A, 124
Sebastiani, Paola, 110–11
selective breeding, 156
self-healing, rapid
    photobiomodulation and,
        106–7
    possibilities of, 105
senescence, 108–9
senolytic drugs, 115–16
senses, augmentation of, 90
shape, enhancement of, *40*, 41,
    43–59
Shoichet, Molly, 38–39
sickle cell anemia, 134–36
Silly Putty, 63
Simon, Joe, 5
Singla, Dinender, 66
single nucleotide polymorphisms
    (SNPs), 78–79, 125
Slack, Ruth, 171–72
soft environment, 155, 157
space
    adaptation to, 118
    as harsh environment, 158

life in, 153

space shuttle, 160

Spider-Man, 61

sports
    doping in, 48–49, 170–71
    enhanced abilities in, 29–30
    superhuman performances in,
        13, 23–24

Star-Spangled Avenger, 5

stem cell delivery, 38

stem cells
    as controversial, 63–64
    fake treatments, 25
    forming eyes from, 65–66
    grafting of, 82–84
    as medical breakthrough, 61
    in muscle tissue, 65
    overview of, 64
    as panacea, 63–64

steroids
    effects of withdrawal from,
        70–71
    effects on cellular memory,
        72–73

Stieglitz, Thomas, 36

Stott, Nicole
    on return from space, 159–60
    on risks of space environment,
        183
    on superheroes, 181–82

Strange, Kevin, 107

strength training
    cellular memory after, 70–73
    response to, 49

stress, responses to, 47–48

super soldier
    procedure, *6, 8, 40*

steps in the making of, 122–*123*

Super Soldier Serum
    failure of, 7
    modern analogues of, 18
    science behind, 9

superheroes
    athletes as, 18, 23–24
    in mythology, 13
    Simon Whitfield on, 2
    see also heroes

superhumans
    bioengineering and, *173*
    creation of, 13
    in mythology, 13

Superman, 103–4

superpower, 186

synthetic biology, 18, 149–50

synthetic DNA, 19–20, 58, 119, 149
    see also DNA (deoxyribonu-
        cleic acid)

Szostak, Jack W., 109

T

Talens, 125

tardigrade DNA, 165

technology, changing human
    capacity through, 122–*123*

telomerase, 109

telomeres, 109–10, 161

testosterone, 50, 69–70

Thor, 62

thymine, 53, 56

Till, Jim, 64

transcranial direct current
    stimulation, 34, 87, 97

transcriptional RNA, 56

transfer RNA, 56

transhumanism, 34

transmutation, 23

trans-species implants, 170

TTN, <u>124</u>

twin studies, 160–61

U

ulnar ligament reconstruction
     surgery, 46

Ultron, 97–98, 177

uracil, 56

uracil-DNA glycosylase, 164

urea, 112–13

V

van Leeuwenhoek, Antony, 148

variome, 142

Venter, Craig, 19–20, 58, 122–23

verbal intelligence, 79

viral vectors, 69, 126, 128

The Virtues of Captain America:
     Modern-Day Lessons on
     Character from a World War II
     Superhero, 28

vision, restoration of, 90

Vita-Ray, *8*, 9, 18

W

water, in body composition, 47

Watson, James, 52

Weber, Doug, 86

White, Mark D., 28

Whitfield, Simon, 1–3, 23

Wilkins, Maurice, 52

Wilson, Daniel H., 37–38, 100

Wilson, James, 131

Windrem, Martha, 83

wingspan, 17–18

Wolf, David, 158

Wolpaw, Jon, 85–86

Wolverine, 104–5

Wolverine: Origin, 104

wood frog (*Rana sylvatica*),
     112–113

Wu, Jun, 84

X

xeno nucleic acid. *see* XNA (xeno
     nucleic acid)

xenobiology, 150–52

X-Men, 132

XNA (xeno nucleic acid), 19,
     149–50

Z

Zimmer, Carl, 128

zinc-finger nucleosis, 125

Zola, Arnim, 87–88

E. Paul Zehr, Ph.D. (neuroscience), is an award-winning science communicator, professor, author, and martial artist. He is the director of the Centre for Biomedical Research and heads the Rehabilitation Neuroscience Laboratory at the University of Victoria and is part of the International Collaboration on Repair Discoveries (ICORD) in Vancouver. His previous books — *Becoming Batman*, *Inventing Iron Man*, and *Project Superhero* — use superheroes as metaphors to explore the science of human potential. He writes for *Psychology Today* and *Scientific American*, among other publications. You can find him on the web at zehr.ca and on social media @E_PaulZehr.

At ECW Press, we want you to enjoy this book in whatever format you like, whenever you like. Leave your print book at home and take the eBook to go! Purchase the print edition and receive the eBook free. Just send an email to ebook@ecwpress.com and include:

- the book title
- the name of the store where you purchased it
- your receipt number
- your preference of file type: PDF or ePub?

A real person will respond to your email with your eBook attached. And thanks for supporting an independently owned Canadian publisher with your purchase!